Machine Learning and Biometrics

Machine Learning and Biometrics

Edited by:

Adele Kuzmiakova

ARCLER
PRESS

www.arclerpress.com

Machine Learning and Biometrics

Adele Kuzmiakova

Arcler Press
224 Shoreacres Road
Burlington, ON L7L 2H2
Canada
www.arclerpress.com
Email: orders@arclereducation.com

ISBN: 978-1-77469-104-5 (Hardcover)

ABOUT THE EDITOR

Adele Kuzmiakova is a computational engineer focusing on solving problems in machine learning, deep learning, and computer vision. Adele attended Cornell University in New York, United States for her undergraduate studies. She studied engineering with a focus on applied math. While at Cornell, she developed close relationships with professors, which enabled her to get involved in academic research to get hands-on experience with solving computational problems. She was also selected to be Accel Roundtable on Entrepreneurship Education (REE) Fellow at Stanford University and spent 3 months working on entrepreneurship projects to get a taste of entrepreneurship and high-growth ventures in engineering and life sciences. The program culminated in giving a presentation on the startup technology and was judged by Stanford faculty and entrepreneurship experts in Silicon Valley. After graduating from Cornell, Adele worked as a data scientist at Swiss Federal Institute of Technology in Lausanne, Switzerland where she focused on developing algorithms and graphical models to analyze chemical pathways in the atmosphere. Adele also pursued graduate studies at Stanford University in the United States where she entered as a recipient of American Association of University Women International Fellowship. The Fellowship enabled her to focus on tackling important research problems in machine learning and computer vision. Some research problems she worked on at Stanford include detecting air pollution from outdoor public webcam images. Specifically, she modified and set up a variety of pre-trained architectures, such as DehazeNet, VGG, and ResNet, on public webcam images to evaluate their ability to predict air quality based on the degree of haze on pictures. Other deep learning problems Adele worked on include investigating the promise of second-order optimizers in deep learning and using neural networks to predict sequences of data in energy consumption. Adele also places an emphasis on continual education and served as a Student Leader in PyTorch scholarship challenge organized by Udacity. Her roles as the Student Leader were helping students debug their code to train neural networks with PyTorch and providing mentorship on technical and career aspects. Her hobbies include skiing, playing tennis, cooking, and meeting new people.

TABLE OF CONTENTS

LIST OF FIGURES

LIST OF TABLES

LIST OF ABBREVIATIONS

AFAS	Automatic Fingerprint Authentication System
AFIS	Automatic Fingerprint Identification/Verification System
AI	Artificial Intelligence
ANFIS	Adaptive Neuro-Fuzzy Systems
ANN	Artificial Neural Networks
ANNMLP	Multi-Layer Perceptron For Artificial Neural Networks
BA	Biometric Authentication
BFS	Brute Force Search
BGP	Binary Gabor Pattern
BSIF	Binarized Statistical Image Features
CNN	Convolutional Neural Networks
CTC	Clock/Counter Time Chip
DCT	Discrete Cosine Transform
DPW	Dynamic Plane Wrapping
DR	Diabetic Retinopathy
DTW	Dynamic Time Warping
DWT	Discrete Wavelet Transform
EER	Equivalent Error Rate
EER	Equal Error Rate
ELM	Extend Learning Model
ERR	Equal Error Rate
ES	Edge Sharpness
EVM	Eulerian Video Magnification
FAR	False Acceptance Rate
FRR	False Rejection Rate
GA	Genetic Algorithm
GLCM	Gray Level Co-Occurrence Matrix

HCV	Hierarchical Visual Codebook-Based
HMM	Hidden Markov Model
HOG	Histogram of Oriented Gradients
ICA	Independent Component Analysis
IPT	Image Processing Toolbox
IR	Infrared
KIT	Key Interval Time
KNN	K-Nearest Neighboring
LBP	Local Binary Pattern
LDA	Linear Discriminant Analysis
LPQ	Local Phase Quantization
MAE	Mean Absolute Error
MAPE	Mean Absolute Percentage Error
MBLTP	Multi-Block Local Ternary Patterns
MLP	Multilayer Perceptron
NB	Naive Bayes
NN	Neural Network
PAD	Presentation Attack Detection
PDTW	Probabilistic Dynamic Time Wrapping
PSO	Particle Swarm Optimization
RBF	Radial Basis Function
RF	Random Forest
RMSE	Root Mean Squared Error
RNN	Recurrent Neural Network
ROC	Receiver Operating Characteristic
RTC	Real-Time Clock
SIFT	Scale-Invariant Feature Transform
SMO	Sequential Minimum Optimization
STFT	Short-Time Fourier Transform
SURF	Speeded Up Robust Features
SVM	Support Vector Machine
TSE	Taylor Series Expansion
TV	Total Variation
UCN	Unconstrained Cohort Normalization

UNHC	United Nations High Commission
VAPE	Variance of Absolute Percentage Error
WLD	Weber Local Descriptor

PREFACE

This textbook provides a comprehensive outline of machine learning designs for biometric and perceptual tasks. The book synthesizes cutting-edge research on applications of CNN (convolutional neural networks) in fingerprint, iris, face, and vascular biometric systems, and soft biometric surveillance systems. MIL-STD-1553 remains the principal network for military avionics integration, despite the advent of higher-speed solutions, such as fiber channel loom. Biometric security issues are also discussed.

Machine learning has been employed to tackle several exciting yet difficult real-world problems, with biometrics being one of the most popular. This book offers several new biometrics and machine learning approaches and recommendations for using machine learning techniques in biometrics. This book covers a variety of "Biometrics" and "Machine Learning" core principles. This book discusses how machine learning may be used to improve biometrics identification performance across an extensive range of biometrics modalities. The book fundamentally explains the following topics:

- Face biometrics is revisited, using insights from neuroimaging and an assessment to popular convolutional neural networks -based face recognition systems.
- This book examines machine learning for state-of-the-art dormant fingerprint, iris detection, and finger–vein. It also discusses ideas for gender classification, gesture-based identification, and tattoo detection using machine learning for soft biometrics. Machine learning for biometrics security is investigated, with approaches for biometrics template protection and liveness detection to protect against fraudulent biometrics samples covered.
- Contributions are presented from a global group of subject experts from industry, academia, and government laboratories.

Chapter 1 gives a comprehensive introduction to biometric and machine learning along with some applications and quantitative assessments, whereas Chapter 2 studies the core concepts of suppression, randomization with classification, and detection using data mining techniques. Chapter 3 presents soft biometrics, their advantages, and performance analysis.

Chapter 4 is dedicated to the recognition by iris and Gabor's demodulation. Chapter 5 introduces the stages of the signature verification system with data acquisition plus a review of signature verification. Chapter 6 covers the application, classification, and types of fingerprint features.

Chapter 7 discusses the introduction and summary of artificial intelligence in biometrics. Chapter 8 is focused on the detection of face, fingerprint and iris presentation attacks.

This authoritative volume will be of great interest to researchers, practitioners, and students involved in related areas of computer vision, pattern recognition, and machine learning, as it provides both an accessible introduction to the practical applications of machine learning in biometrics and comprehensive coverage of the entire spectrum of biometric modalities.

—Author

Chapter 1

Machine Learning for Biometrics

CONTENTS

1.1. INTRODUCTION

We are seeing a significant trend using big data and artificial intelligence algorithms. As a result, these algorithms are getting increasingly more sophisticated and accurate. We also use high-speed WiFi to exchange data in near real-time and make revolutionary lifestyle modifications. Machine learning methods, which are at the heart of artificial intelligence, are designed to build sophisticated solutions and solve current problems in a variety of disciplines, ranging from business, engineering, and pure art to science.

Biometric technology is gaining momentum as a critical technology for internet-age smart setups, ensuring both network and computer security. Modern machine learning methods with state-of-the-art designs may be used to such biometric setups to reach extremely high-level efficiency and become smarter as planned (Viola & Jones, 2004). We investigate various exemplary biometrics in the book, examine significant features of the specimen from relevant biometrics, and define their relevant attributes and descriptions. In different sections, we cover well-managed machine learning techniques and big data, as well as their implications to biometric research.

1.1.1. Biometrics

Biometrics has advanced significantly in current years, with many uses in everyday life. Utilizing various methods, biometric identification identifies persons depending upon biometric features (Aggarwal et al., 2004). Biometrics is divided into two categories based on their features: physical features and psychological attributes. Direct physical proof of discriminated characteristics in biometrics with physical features can be found in their specimens. Biometrics like the palm veins, face, iris, retina, fingerprint, ears, and DNA fall under this group. Discriminative characteristics from parameters with psychological traits, on the other hand, may be retrieved automatically from specimens, and this group includes biometrics like stride, typing rhythm, and sound.

A biometric setup for authentication may be executed and developed in one of two ways: feature-based with manual characteristic extraction or automatic characteristic creation with end-to-end training utilizing a machine learning technique. To decrease the unpredictability and computing complication of innovative characteristics in feature-based biometric setups, the choice of characteristics descriptors and kinds, as well as the subsequent classification design, becomes critical (Turk & Pentland, 1991). Every characteristics descriptor excels at describing a certain pattern kind. The

Gabor filter, for instance, offers higher frequency and directional selectivity, allowing it to be utilized to perform the time-frequency evaluation on input pictures. LBP and its variations are texture coding operators that are usually resistant to changes in lighting and facial expression in pictures. As a result, it's important to pick the correct characteristic for the right uses. A feature-based setup's effectiveness is mostly dependent on the abilities of human specialists, which typically results in limited generalization for variances in input data. Current automatic characteristics generation-based techniques, like deep learning, may be a good way to cope with these issues. The feature choice and classifier components of this approach are trained simultaneously with a huge quantity of data, and it performs better than a feature-based setup. One downside of this technique is that it generally involves a large number of variables that must be trained over a lengthy period.

Biometric setups with high precision and minimal computing complexity are ideal for security applications. A trustworthy biometric setup must also generalize well enough for unseen specimens and be incredibly durable to a variety of difficulties, such as structural enhancement, light variation, noise, and intra-class variance. We ought to build a setup that can withstand many sorts of assaults, like counterfeiting, for real-world security use. Multimodal biometrics (Ahonen et al., 2006) has gotten a lot of attention recently as a way to build a more secure setup, and it's become a hot study issue. A multi-functional setup, for instance, with the input of fingerprint pictures & finger veins, provides broader application, greater dependability, and stronger safety than a unimodal setup in actual use. Biometrics is fundamentally an issue of categorization. It may be thought of as a multi-class issue in its most basic form (e.g., if no previous individual identity is provided). Alternatively, in the verification set, in which a prior on the personal identification is known, a binary classifier that distinguishes among real users and frauds might be considered. A generic workflow comprises a classification stage and a data representation stage in this application (He et al., 2005). If the biometrics under consideration has a significant intra-class variation, the learning from experiences approach may be used. Statistical learning is an excellent methodology for both data representation (e.g., using data-driven feature extraction methods) and strategic planning.

1.1.2. Machine Learning

Machine learning focuses on optimizing system parameters based on design to fulfill assessment criteria using a collection of training instances (Argyriou et al., 2008). We frequently employ statistical approaches to enable computers to "learn." Based on the utilization, we can utilize the resultant system to forecast the categorization of different datasets, predict position in the feature space, or produce fake instances after the learning objective has been accomplished. Un-supervised learning, supervised learning, and reinforcement learning are the three main kinds of machine learning methods.

The training datasets for supervised learning issues consist of cases of the input vectors and their matching outcome vectors (Etemad & Chellappa, 1997). The issues are classified into categorization or pattern identification when the target vectors are classified, and prediction once the target vectors are genuine. For every input vector, deterioration or distance factors are constructed among the recent outcome vector and the desired vector, and an improvement is conducted to reduce the overall deterioration of training samples. We anticipate the system to react properly although if the unknown dataset is provided to the trained setup since we taught it with known input and target pairings.

There are no objectives in unsupervised learning tasks, thus the training dataset is just a set of input vectors. The objective of unsupervised learning is to find "important statistical design" in data autonomously. Unsupervised learning is also known as hidden knowledge discovery from models, and several clustering techniques are standard models.

Learning how to respond in a particular scenario for certain incentive or punishment signals is known as reinforcement learning (Ando et al., 2005). A condition for the present position is specified in this form of learning, as well as the environment, generally, a criteria function assesses the present situation to create an appropriate incentive or punishment action via a set of principles. It learns from criticism rather than having specific objective values.

Deep learning is a subset of machine learning and has shown to be effective in areas such as differentiation, identification, key point estimate, categorization, and activity categorization (Argyriou et al., 2008). It usually has a wide range of parameters as well as numerous nonlinear films. CNN (Convolution neural networks) for automatic feature extraction and RNN (recurrent neural networks) for series prediction are two prominent deep learning designs. They have demonstrated remarkable improvement in machine vision, speech recognition, audio identification, voice recognition, social media screening, bioinformatics, and translation software. Furthermore, generative models, such as nonlinear encoders and

GANs (generative adversarial networks) are gaining popularity due to their capacity to generate synthetic samples.

We examine an optimization technique to regularized feature choice depending on a 1-norm penalty term and certain current improvements in learning features from raw data. We then provide a feature learning technique for multi-functional apps that enables us to identify feature groups about various activities autonomously. Such methods apply to a variety of biometric use since they are better suited to biometric data that have little or no effectiveness and uniqueness reliability, such as gait, face, or signature biometrics, in which the learning from instances framework has been demonstrated to be efficient (Jain et al., 2004). The focus of this article is on two-dimensional face biometry. The first depiction in this situation may be found in computer vision research and would not be the subject of this section. Rather, we'll look at how feature choice can be used as an automated technique for generating concise representations of information, which can help with both categorization efficiency and time complication. Furthermore, several intra-class informative picture representations are frequently highly complex; therefore dimensionality reduction methods are common throughout this application space (Belhumeur et al., 1997; Destrero et al., 2007).

The choice of features is a new method of collecting useful data from samples and has been widely investigated in the single, functional situation. Various approaches have emerged for its usefulness (Weston et al., 2003; Torralba et al., 2004). Regularization approaches are another intriguing approach to cope with the choice of features in the learning through experiences paradigm. The theoretical foundation for this technique may be found by Chen et al. (2001). Destrero et al. (2009) provide more information on the technique mentioned in this section.

Whereas the challenge of learning the relevant data embedded in a specified data depiction has been widely explored in the single-task unsupervised learning and supervised learning settings, there has been little research in the multifunctional supervised learning environment (Ando et al., 2005). The challenge of learning a basic design shared by a group of supervised jobs is examined in this section. This is especially beneficial for improving classification capability when the activities are numerous however the available data is limited. Knowledge of the basic design can also make learning additional challenges easier. Argyriou et al. (2008)

provide more information on the claimed technique. We examine the more basic issue of feature learning in the situation of multi-function learning, which is the assessment of a batch of new features generated from several input parameters. The choice of feature may be viewed as a specific case in this approach.

It's important to note that the regularization variable handles the complication of the solution in both situations: in single-function the choice of feature, it handles the nonlinearity of the solution, whereas, in multi-function feature learning, it handles a range of features familiar to all of the activities learned.

1.2. CONDITION OF THE ART ON FACE BIOMETRICS

Face biometry has benefitted from machine learning improvements more than other biometrics. Face biometry focuses mostly on the *preprocessing phase*. Apart from that, face identification and face detection features have been thoroughly investigated.

Facial recognition represents a binary categorization problem. Here we want to classify the data into facial or non-facial (–ve) categories. Machine learning techniques, such as support vector machines or neural networks were used in early attempts at facial recognition (Sung & Poggio, 1998). The dataset was chosen to represent system's requirements. View-based methods were frequently utilized to represent diversity in perspectives. Local regions are more suited to cope with pose variations and malformations. The application of overfitted dictionaries of local features capable of capturing local designs and moderate geometries, as well as feature choice techniques that decrease the duplication of the first description, is the latest trend popularized by the seminal work of Jones & Viola (2004). These approaches are famous since they are efficient and can be used successfully in real-life situations.

1.3. LATEST ADVANCES IN MACHINE LEARNING

This chapter discusses the latest advancements in computational learning theory for learning dataset representations. Such approaches are suitable for dealing with picture dataset redundancy and the complication of multi-class issues autonomously.

1.3.1. Learning Characteristics for Binary Categorization

We look at how to choose a collection of characteristics that are expressive of a binary categorization task. We examine a currently suggested regularization approach for learning features from a set of instances from two separate classes in this chapter. This technique is dependent on Lasso technique (Tibshirani, 1996).

1.3.1.1. Issue Formulation

We discuss a binary categorization issue in which we are assigned an "m" element training set. A glossary of "d" characteristics is used to describe every instance. Using feature choice, we seek a concise description of data input for the issue of concern. Every characteristic in our solution is assigned a weight; characteristics with greater than zero weights are significant for modeling the variety among the two classes under concern.

We look at the scenario when the incoming and outgoing data are linearly related. The solution to the given linear setups of equations may be used to formulate the issue:

$$y = Xa \qquad (1)$$

wherein $X = \{x_{ij}\}, i = 1,...,m; j = 1,...,d$ is the $(m \times d)$ features matrix acquired showing the training dataset of "m" inputs x_i with a glossary of "d" characteristics. The output tags are stored in the "m" vector $y = (y_1,..., y_m)$: because we are using a binary categorization setup, all information is linked with a label $y_i \in \{-1,1\}$. Every item is connected with one feature and essentially indicates the significance of the feature in deciding the affiliation of a particular feature vector to one of the two categories, $a = (a_1,..., a_d)$ is the vector of unfamiliar weights, every item is linked with one characteristic (Sundararajan & Woodard, 2018).

We assume the sizes of X to be enormous in the utilization areas, therefore the typical techniques for resolving the algebraic setup (10.1) are impracticable. Furthermore, the typical amount of features "d" is substantially greater than the training set's size "m," indicating that the network is significantly inadequate. We can also have to cope with excitement to play among feature vectors that create significant sick-conditioning due to the duplication of the feature dataset. These problems need regularization, which may be avoided by converting issue (10.1) to a penalized least-squares issue.

A quadratic penalized, generally the two-norm of the vector a: $\|a\|_2^2 = \sum_j a_j^2$, is used in traditional regularization, like the familiar ridge regression (also known as Tikhonov regularization) (Palaniappan & Mandic, 2007). The solution of the penalized lowest-squares issue would generally produce a vector including all weights a_j distinct from zero, but these quadratic charges don't offer feature choice. That's why, in the latest research, the use of sequence-enforcing charges to substitute quadratic penalties has been recommended. It indicates that the existence of zero weights in vector "a" would be enforced by the fine. The one-norm of "a" is the sole convex penalty between these zero-enforcing charges, allowing for viable methods for the higher-dimensional dataset. As a result, we analyze the following issue, which is known as Lasso regression (Tibshirani, 1996):

$$a_L = \arg\min_a \{\|y - Xa\|^2 + 2\tau \|a\|_1\}, \tag{2}$$

wherein $\|a\|_1 = \sum_j |a_j|$ is the one-norm of "a" and "τ" is greater than zero is a regularization factor regulating the equilibrium among the data mismatch and the fine. This variable also modifies the level of heterogeneity (number of genuine zero weights) of the vector" a" in feature choice issues.

1.3.1.2. Learning Method

Because of the one-norm penalty in issue (2), Lasso solutions' reliance on y is non-linear. As a result, computing one-norm penalized solutions is much more complex as compared to computing two-norm penalized solutions. We use a simple iterative method suggested by Daubechies et al. (2004) to resolve (10.2):

$$a_L^{(n+1)} = S_\tau[a_L^{(n)} + X^\top(y - Xa_L^{(n)})] \quad n = 0, 1, \tag{3}$$

employing random preliminary vector $a_L^{(0)}$, wherein S_τ is the following "soft-threshold"

$$(S_\tau h)_j = \begin{cases} h_j - \tau \operatorname{sign}(h_j) & \text{if } |h_j| \geq \tau \\ 0 & \text{otherwise} \end{cases}.$$

That technique is referred to as the Landweber iteration in the absence of a soft threshold (τ= 0), that intersects to the generalized solution (minimal-norm lowest-squares solution) of (10.1). Daubechies et al. (2004) showed that the soft threshold Landweber technique (10.3) converges to a minimizer of (10.2) if the norm of the medium "X" is renormalized to a value rigorously less than one.

Because experimental data suggests that the startup vector is unimportant, we always begin the weight vector "a" with zeros: $a^{(0)} = \mathbf{0}^{\top}$. The iterative process' halting rule is depending upon comparing the solution acquired at the nth iteration $a^{(n)}$ with the prior one, and it is connected to the sustainability of the solution attained. The pseudo-code of the iterative method is reported in the method in Table 1.1.

1.3.1.3. Feature Choice for Large Issues

In a biometric context, numerous issues that may arise. As a result, implementing the iterative process given in Equation (3) can't be possible across all PCs: matrix multiplication must be properly designed such that the whole matrix "X" is not kept in primary memory.

Table 1.1. Algorithm 1: Binary Feature Choice

Input: training dataset (in the suitable demonstration)

$\{(x_i, y_i)\}_{i=1}^{m}\ y_i = \{-1, 1\}, x_i = (X_{i1}, \ldots, X_{id})^{\top}\ X = (x_1, \ldots, x_m)^{\top}$

Parameters: regularization factor τ

Output: the light weights vector a **Initialization: for all** j $a_j = 0$; **end for** standardize medium "X" (see text) **as** outlet circumstances are met **do** modernize

$a = a + X^{\top}(y - Xa)$

for i is equal to1,...,m **do**

if $|a_i \geq \tau|$ **then**

$a_i = a_i = \tau \text{sign}(a_i)$

else

a_i is equal to 0;

end if

end for

end while

We design a technique depending upon resampling the feature dataset and getting numerous smaller issues: we create "S" feature subsets every time removing with replacement "p" features from the basic set of size "d" ($p << d$), and afterward we acquire "S" smaller linear sub-issues of the type: Xs as = y for s = 1,..., S, where X_s is a submatrix of X with columns corresponding to the characteristics in s; a_s is calculated correspondingly.

We note that the subset dimension must be large enough to be expressive but little enough to manage the matrix effectively; hence, we choose subsets with 10% of the initial feature set dimension. We use the binomial distribution to determine the number of sub-issues "S" and approximate how many separations are required such that every feature gets retrieved at most 10 times with a high possibility (Destrero et al., 2007).

Following the construction of the "S" sub-issues, we search for the "S" solutions using "S" iterative techniques, as shown in Equation (10.3). We are left with "S" redundant sets of features after the procedure. The finalized set is created by selecting the characteristics that have been picked in every one of the subsets.

The two matrix–vector multiplications create the computational complication of the iterative method (10.3), making it O (md) for every repetition and a constant τ. We choose O because the iterations number "I" is not insignificant (mdI). This must be performed several times as the range of models we are evaluating throughout the model selection stage.

It's important to note that the preprocessed form of the technique is appropriate for parallel processing, allowing for savings in calculation time inverse proportional to the range of processors employed (Donoho, 2000). If just a single processor is accessible, the model choice step must be carefully designed to keep the minimum calculation cost of training stage computing In our studies, we use two techniques: (i) choosing τ that contains a constant amount of 0s in the solution (or, comparably, which chooses a specified range of attributes) in about "I" iterations, and (ii) to choosing τ depending upon generalization capability, utilizing cross-validation: we choose "τs" that contribute to the categorization rates under a certain limit and then choose between them, the value giving the small range of features. Equation (10.1) is utilized as a classifier in this case (Edwards et al., 1999).

1.3.1.4. Assessment techniques

We end this chapter by describing how we assess the quality of the acquired subset of characteristics. We start by recalling that we are looking for features that will help us to construct a good classifier. As a result, the generalization efficiency of a classifier trained on a data set with the specified features is assessed.

We employ an asymmetric, SVM-like (support vector machine) categorization method. This is utilized to construct the final recognition and verification modules. Receiver operating characteristic (ROC) curves describe efficiency, with the support vector machine offsetting value "b" being varied. ROC contains the strike rate equivalent to 0.5% false-positive results (which is especially essential for recognition issues with a large range of negativity) and the equivalent error rate (EER) (Cherkassky & Ma, 2009).

1.3.2. Learning Shared Depictions for Several Activities

In this part, we look at how to learn several predictions or categorization activities at the same time. We look at a technique for learning a collection of characteristics that are common to several jobs. This is an issue that has attracted the interest of many researchers, and its usefulness was explored by Torralba et al. (2004) in the context of object recognition. Images of various things can have several characteristics which vary from the pixel images depiction. Such technique may be useful in the biometry area when coping with the multi-class issue of facial identification: various persons can share numerous characteristics, therefore understanding such general aspects may help with various recognition activities.

1.3.2.1. The Formulation of Issue

"T" supervised learning assignments are assigned to us. For each $t \in N_T$, the related activity is recognized by a function ft: $R^d \to R$ (such as a margin / repressors classifier). For every activity "t," we are provided with a collection of "m" output or input samples, that we organize in an (m × d) matrix X_t and an "m" vector y_t for t = 1,..., T, similarly to the preceding chapter.

We want to utilize the samples given to find specific connections between the activities. The variables f_t can be expressed as a linear combination of certain feature tasks, which is our working supposition. For the sake of convenience, we only examine linear identical features, all of which are expressed through a vector $u_i \in R^d$ – non-linear features are covered in (Argyriou et al., 2008). In addition, we suppose that the vectors u_i are orthogonal, thus we simply analyze up to d of every vector.

The activity functions may be expressed as $f_t(x) = \langle a_t, U^\top x \rangle, x \in \mathbb{R}^d$, where $a_t = (a_{t1},...,a_{td})$ is the vector of regression coefficients for the tth activity, and $U \in \mathbf{O}_d$ is the matrix whose columns are the vectors u_i.

Our supposition that the activities share a "limited" set of characteristics implies that matrix A has "several" rows that are all equivalent to zero, and therefore no activity would utilize the associated features (columns of matrix U).

The optimization issue is solved by our approach for multi-function feature learning

$$\min\left\{\mathcal{E}(A,U) : U \in \mathbf{O}^d,\ A \in \mathbb{R}^{d\times T}\right\}, \tag{4}$$

$$\mathcal{E}(A,U) = \sum_{t=1}^{T} \|y_t - X_t U a_t\|^2 + \gamma \|A\|_{2,1}^2, \tag{5}$$

wherein γ is greater than zero is a regularization factor.

In the 1-term in Equation (5), the square loss might be exchanged with a loss task

L: R×R → R+ convex in the 2nd argument. The 2nd term is a normalization term, which discourages matrix A's (first and second) norm. It's calculated by calculating the second norms of the (across activities) rows a^i (correlating to feature (i) and then the first norm of the vector ($\|a^1\|_2, \ldots, \|a^d\|_2$). The magnitudes of the vector's elements reflect the importance of every feature.

If the medium "U" does not educate and we set up $U = I_d \times d$, issue (4) chooses a "limited" parameter set, generally across the activities. In addition, if T = 1, function (5) simplifies to the first norm normalization issue mentioned in the previous chapter (Pontile et al., 2007).

The preceding (first and second)-norm promotes a limited range of nonzero rows in matrix A, assuring that just certain general characteristics are acquired across activities. The amount of features learned is dependent upon the amount of parameter γ, and it would unlikely to increase with γ.

1.3.2.2. Equivalent Convex Problem

Issue (4) is difficult to solve for two reasons. Firstly, it is a non-convex issue, despite being convex in every one of the parameters A and U individually. Secondly, because the regularizer $\|A\|_{2,1}^2$ isn't uniform, the optimization issue becomes harder to resolve (Bicego et al., 2006).

Luckily, issue (10.4) may be converted into a convex version.

To explain this conclusion, we explain a function that replaces the square loss with a more generic loss L for any $W \in \mathrm{R}^{d\times T}$ with columns w_t and $D \in \mathbf{S}^d$, ++.

$$\mathcal{R}(W,D) = \sum_{t=1}^{T}\sum_{i=1}^{m} L(y_{ti}, \langle w_t, x_{ti}\rangle) + \gamma \operatorname{trace}(D^{-1} W W^{\top}) \tag{6}$$

Therefore, It is likely to show that issue (4) is equal to the convex optimization issue.

$$\inf\left\{\mathcal{R}(W, D) \,:\, W \in \mathbb{R}^{d\times T},\ D \in \mathbf{S}^{d}_{++},\ \operatorname{trace}(D) \le 1\right\}. \qquad (7)$$

Particularly, every minimizing series of issues (7) corresponds to a minimizer of the issue (4) & (5). Furthermore, the formula connects the solutions ($W\hat{}$, $D\hat{}$) and ($A\hat{}$, $U\hat{}$) of issues (7) & (4), accordingly.

$$(\hat{W}, \hat{D}) = \left(\hat{U}\hat{A},\ \hat{U}\operatorname{Diag}\left(\frac{\|\hat{a}^i\|_2}{\|\hat{A}\|_{2,1}}\right)_{i=1}^{d}\hat{U}^{\top}\right)$$

Observe that we restrict the tracing of matrix D above from the issue (7) as otherwise, the best solution will be to just set $D = \infty$ and merely minimize the empirical error term in the right-hand side of Equation (6).

1.3.2.3. Learning Method

Now we look at a method for addressing the convex optimization issue (10.7). The procedure minimizes a disturbance of the optimal function (10.6) by adding a disturbance "I" to the matrix WW, which appears in the 2nd term on the right-hand side of (10.7), where "I" is greater than zero and "I" is the identity matrix. This disturbance assures that D remains nonsingular and that the infimum over D is often achieved (Pires et al., 2021).

The two phases of the Method for addressing this unsettled issue are now described in Table 1.2. We maintain the D constant and minimize over W in the first phase. Because the regularizer detaches when D is constant, this phase may be performed separately across activities. Introducing additional parameters for $D^{-\frac{1}{2}}w_t$ results in a typical 2nd norm regularization issue for every activity with the similar kernel $K(x, x') = x^{\top}Dx'$ for every job. In the 2nd phase, we adjust matrix W and reduce D to the smallest possible value.

Table 1.2. Algorithm 2: Multi-Functional Feature Learning

Input: training sets $\{(x_{ti}, y_{ti})\}_{i=1}^{m}, t \in \mathbb{N}_T$

Parameters: regularization parameter γ, tolerances ε, tol

Output: $d \times d$ matrix D, $d \times T$ regression matrix $W = [w_1, \ldots, w_T]$

Initialization: set $D = \frac{I_d}{d}$

while $\|W - W_{prev}\| > tol$ **do**

  for $t = 1, \ldots, T$ **do**

    compute $w_t = \operatorname{argmin}\left\{\sum_{i=1}^{m} L(y_{ti}, \langle w, x_{ti}\rangle) + \gamma\langle w, D^{-1}w\rangle : w \in \mathbb{R}^d\right\}$

  end for

  set $D = \frac{(WW^{\top}+\varepsilon I_d)^{\frac{1}{2}}}{\operatorname{trace}(WW^{\top}+\varepsilon I_d)^{\frac{1}{2}}}$

end while

The equation may be used to prove that partial reduction in the case of D has a closed-form resolution.

$$D_{\varepsilon}(W) = \frac{(WW^{\top} + \varepsilon I_d)^{\frac{1}{2}}}{\operatorname{trace}(WW^{\top} + \varepsilon I_d)^{\frac{1}{2}}}. \quad (8)$$

The method may be thought of as switching between unsupervised & supervised stages. Utilizing a similar depiction across activities, we understand task-specific functions (particularly the vectors w_t) in the first phase. Since D encompasses the features u_i, the feature depiction remains constant. The regression functions are constant in the unsupervised stage, and we learn the general depiction.

It is demonstrated in Argyriou et al. (2008) that the method converges to a solution to the relevant disturbing issue for any I>0. Furthermore, every limiting point in the series of these solutions fixes the issue (10.7), as the saying goes.

We also make a few observations on an alternate formulation for the issue. We have a regularization issue in W if we replace W= 0 in Equation (10.7), where the regularization term is the square of the trace norm of matrix W, i.e., the total of singular values of W. The trace norm is the convex enclosure of rank (W) in the unit ball, as represented in (Fazel et al., 2001). We should also mention that a similar topic was investigated from the perspective of joint filtering (Srebro et al., 2005).

1.4. APPLICATIONS

This chapter explains how we use statistical learning approaches to tackle problems in biometry. We illustrate how feature choice can be used for both facial recognition and facial identification. We then give numerical simulations and explore how multi-functional feature learning can be used to detect issues.

1.4.1. The Peculiarity of Facial Features

A significant degree of duplication is seen in image characteristics. The curse of dimensionality can impair the quality of results if we work with an unbalanced training set. While the redundancy in the training set doesn't undermine a classifier's generalization ability (Donoho, 2000), it can have an impact on the classifier's performance.

Dealing with information duplication through the choice of feature entails picking one or few delegates to show the other parts of every set of linked characteristics. In contrast to other application areas (like microarray statistical analysis), typical image-related issues allow for the selection of an arbitrary delegation for the linked feature datasets. The look data that image features contain, that is similar to other participants of their group, is more essential than the set itself (Guyon & Elisseeff, 2003).

Facial characteristics can be linked not just because of all-natural photos' inherent short-range association or because the selected explanation is repetitive, and because of interdependence associated with the face class (which contains numerous appearances of identical patterns at various places). Comparison based on the selected depiction provides redundant explanations, but association based on the class of interest can convey valuable data about its characteristics. A feature selection strategy must address both of such requirements; however, a minimal number of features is generally chosen for statistical reasons (Heiselet et al., 2001).

1.4.2. Face Recognition

In this part, we concentrate on facial feature choice before briefly reviewing the construction of an effective facial detector, beginning with an image depiction dependent upon rectangle features which have shown to be a useful beginning point for several object identification issues (Viola & Jones, 2004).

The raw data we're looking at are image patches of size $N \times N$ (N = 19). For every picture or image patch in the examination, we calculate the overall positions, sizes, and aspect ratios of rectangle features. As a result, we calculate around 64,000 characteristics per patch or image. Now let's look at how we use the iterative process outlined in Section 13.3 to pick facial characteristics.

We have 4000 training datasets, 2000 actual data, and 3400 test data, equally divided among positive (facial images) and negative (non-facial images). The linear setup we create for feature selection is rather large, with a data matrix A of 4000 × 64,000 (because every entry is accumulated in solitary accuracy, the overall space needed for matrix A is around one Gigabyte) (Fuentes et al., 2010).

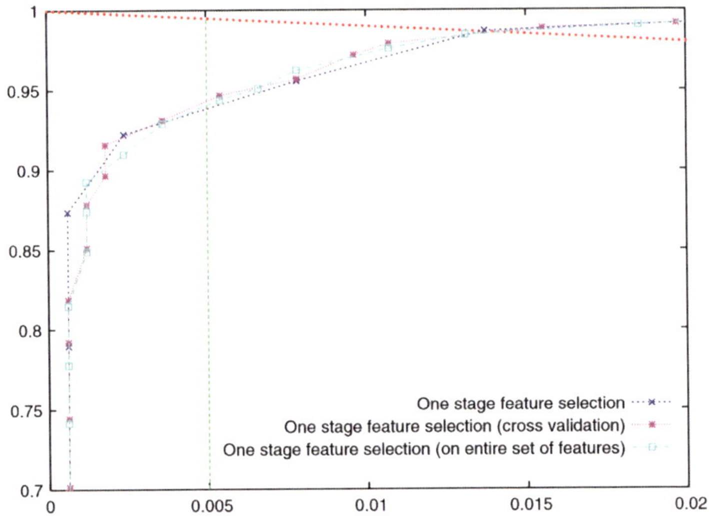

Figure 1.1. Relationship among (1) a straightforward solution of the linear issue (–o–) and (2) the resampling approach (with two various variable selection, setup of sparsity level (–x–) or by cross-validation (–*–)).

Source: *https://link.springer.com/chapter/10.1007/978–1-84882–385–3_10.*

We present an extensive number of experiments that demonstrate the resampling plan's suitability as the size of the data matrix rises (Destrero et al., 2007). Figure 1.1 summarizes the findings achieved without and with the resampling approach, demonstrating that there are not only zero losses but even a modest gain when using the resampling method. We decide τ to retain a constant amount of characteristics when it pertains to model choice since we have to deal with a large number of distinct challenges (such as search solutions with a specified % of zero entries). Figure 1.1 likewise depicts the results achieved using a cross-validation-based model selection (the results are relatable in such 1st phase).

The leftover features S_1 are a fair composite of the initial description, as proven by the categorization results on our testing set, and they preserve all the descriptiveness, reflecting all significant parts of a facial, after such a feature selection phase. Despite this, the range of chosen features remains considerable (around 4500 in our experiments). Rather than picking a value for τ that forces a greater range of zeros (that option resulted in a significant loss in efficiency, check we perform the feature selection method to fixed S_1

in search of a better, sparser solution (Destrero et al., 2007). The latest data matrix is achieved by choosing the columns that relate to S_1 from A, and $f_{S_1}^{(0)}$ is reset to a vector of zeros.

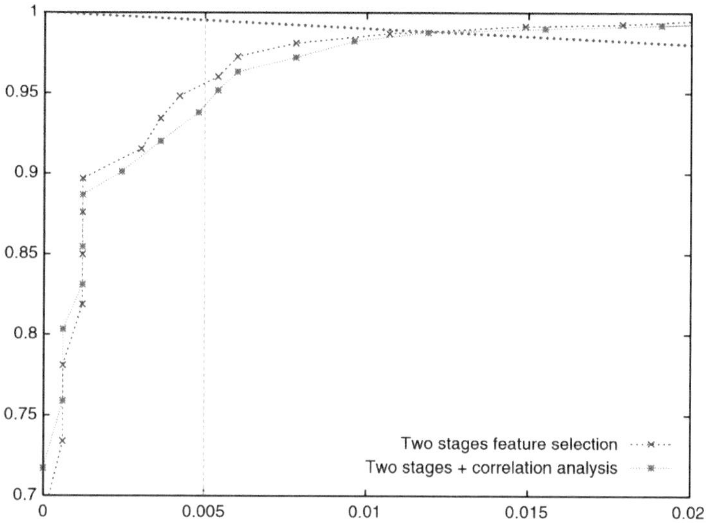

Figure 1.2. Results acquired with a 2-phase assortment process, without (–x–) and with (–*–) the additional relationship analysis).

Source: *https://link.springer.com/book/10.1007/978–1-84882–385–3.*

In this 2nd step, we will utilize cross-validation like for variable tuning technique. When statistically possible, cross-validation includes an explicit account of adaptation and is thus more suitable. Figure 1.2 shows the results of the 2-phase selection method, which yielded a set S_2 of 247 features with no efficiency degradation compared to the 1st phase. Figure 1.2 also displays the categorization results of the feature set S_3 generated by performing basic correlation analysis on S_2 to preserve just one feature among sets with numerous characteristics of a similar kind, strong relationship, and spatially contiguous features (Ullman et al., 2002).

The data for the outcomes we provide here was collected using a surveillance system established in our department and individually labeled as either positive or negative. The identified images represented frontal facial features. We employed the CBCL-CMU frontal faces set of data in a new series of tests. Because of the varied nature of the data sets, the findings achieved in both situations are unique. Faces are only roughly registered and information contains considerable perspective variances when utilizing our data collection. As a result, horizontal symmetry is retained but vertical symmetry is not unique to the set of data (see Figure. 1.3) (Weston et al., 2003).

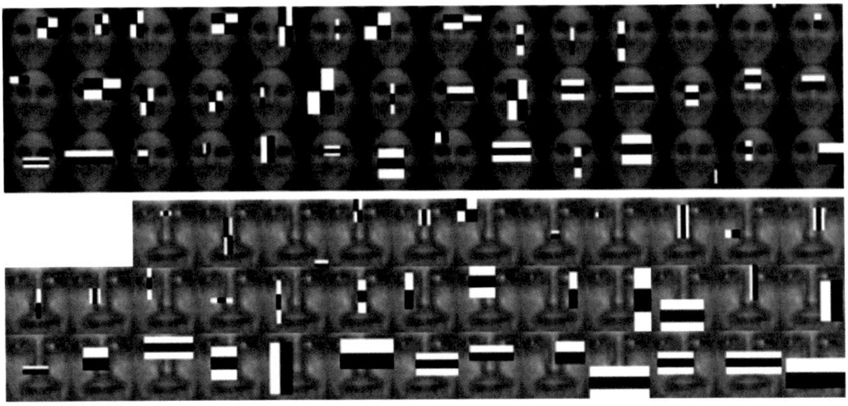

Figure 1.3. Top: the chosen features from the DISI set of data. Bottom: chosen features from the CBCL-CMU set of data.

Source: *https://link.springer.com/book/10.1007/978–1-84882–385–3.*

We wrap up with a description of the finalized face identifier and the results received. More details can be found in Destrero et al., 2009; Ranftl et al., 2017. S3 characteristics are utilized to create a waterfall of tiny SVMs capable of processing video frames in actual time. The picture is analyzed by a sequence of classifiers using the conventional coarse-to-fine method. Every picture patch is the input for the classifiers' waterfall: when the patch failed one test, it is instantly rejected; whether it satisfies all tests, a face is recognized. Every classifier in the cascade is constructed with different features, training a linear SVM on those characteristics, then inserting more features till the validation set achieves the specified efficiency. We specify the minimal strike rate to 99.5% and the max false-positive rate to 50% to ensure that every classifier doesn't overlook faces. Considering a cascade of 10 classifiers, the overall efficiency will be: $FPR = 0.5^{10} – 3 \times 10^{-5}$ and $HR = 0.995^{10}$ □ 0.9 (Viola & Jones, 2004).

In two distinct circumstances, Table 10.3 illustrates its recognition efficiency as a facial identifier in our actual-time setup. The finding acquired on photos taken in controlled environments (individuals were instructed to approach the cameras one at a time) are shown in the first row of the table, whereas the finding acquired on the image obtained in uncontrolled environments are shown in the second row: We physically labeled the incidents that occurred during a five-hour recording of a hectic day; the recording was completely outside from our control and includes scene variations, individuals halting for an undetermined amount of time, and sideways faces. It's important to note that the classifiers have been fine-tuned in real-time to reduce the amount of false-positive (Zhao et al., 2003). The quantity of data processed is enormous, as the detector examines 20,000 patches on every picture or screen.

Table 3. Characteristics of Face Identification Setup

Test data	**Strike rate percentage**	**False-positive rate percentage**
Live video	76	4 10−7
Test pictures	94	0.1

1.4.3. Face Identification

We now look at a facial identification issue that involves distinguishing between instances of the person whose identity has been proclaimed and samples of other individuals. Particularly, we explore a challenging set-up that represents intra-class vs extra-class variability for every person. At a constant scale, we examine a picture description depending upon LBP characteristics (Ahonen et al., 2006).

We suppose that we have an information set of N person I1,..., IN, Let's take a quick look at how we express every personal I:

We begin with a collection of 40 by 40 +ve photos I_p, $p = 1,\dots, P$ of person I and a set of –ve photos I_n, $n = 1,\dots, N$ of other people randomly chosen.

Every neighborhood in picture I_i is depicted by an LBP. We calculate the size of the neighborhood by selecting eight sample sites on a circle with a radius of two. Furthermore, for every picture, we calculate L LBP graphs on rectangular areas of at least 3 by 3 pixels in all places and aspect ratios, therefore the explanation of picture I_i a list of histograms $H_i^1,\dots, H_i^L$. Because the number of feature vectors is quite large, a feature choice method is necessary: we acquired L = 23409 LBP graphs (Argyriou et al., 2008).

Segment 13.2's regularized feature choice is used once more. The linear system is designed to convey every histogram's intra-personal and extra personal variance. We calculate the feature vectors (in the particular instance the rows of the matrix (a) in the same way by comparing correlating LBP histograms (after normalization) for every pair of photos I_A & I_B utilizing the $\chi 2$ distance (Ouamane et al., 2014). Such that, we acquire a feature vector x_i whose components are $\chi 2$ distances for every pair of input data: $x_i = (\chi^2(I_A^1, I_B^1),...,\chi^2(I_A^L, I_B^L))$. We correlate a label $g_i \in \{+1,-1\}$, wherein $g_i = 1$ however both I_A & I_B are positive images, $g_i = -1$ if I_A or I_B is negative.

The range of feature vectors that we can compute for a specified set of instances is enormous (for a set of positive photos of size P we would have $\binom{P}{2}$ positive feature vectors, and the –ve, in general, would be very high). We randomly pick at most 2000 positive and negative feature vectors and create matrix A to achieve stable systems of appropriate size. The vector g is made up of the matching tags g_i, and the unknown's vector f would calculate the significance of every LBP graph for extra and Intra person comparisons. After we've constructed a system for a certain person using the process outlined above, we choose characteristics using the same procedure outlined for face recognition. It's worth noting that we get a various set of distinguishing traits for every person, the ones that best express his or her distinctiveness (Heiselet et al., 2001).

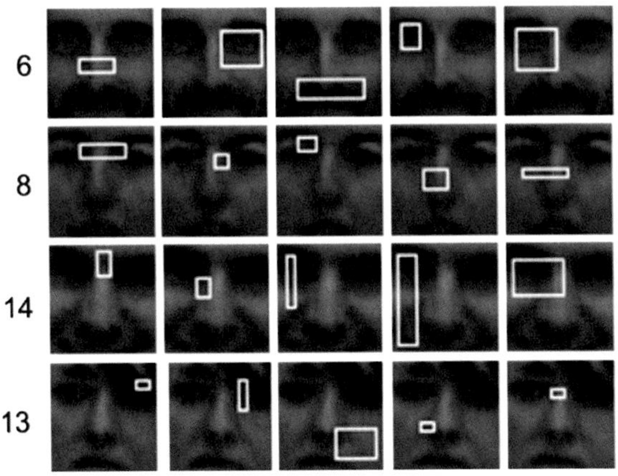

Figure 1.4. Top five characteristics for certain persons: they frequently catch different face features (hair, facial hair, or a clown nose; (bottom row).

Source: *https://link.springer.com/book/10.1007/978–3-642–35326–0.*

We assess a feature set's quality in terms of its generalization ability once more. The facial recognition module automatically gathers and registers the given dataset that we need for face recognition. The experimentation in this section, particularly, is acquired with our monitoring program over many weeks: at the end of two weeks, we individually classified the stored videos (a video is stocked if faces are identified in it) and constructed models for all of the people with a large set of data (Ranftl et al., 2017). Persons who had not previously appeared were kept as negative instances (frauds) for all designs starting in the 3rd week. A sum of sixteen people was involved in the training stage, whereas a sum of 64 people was collected for the testing stage. The top five characteristics retrieved for various people are shown in Figure 1.4. Take note of how every person's characteristics cluster in the most distinguishing parts of the face. The beard, for example, may be seen on individual six. Person thirteenth top two characteristics are focused in which there is a perimeter. In this example, the nose is where the majority of the characteristics were taken (Manresa et al., 2014).

Now let us go through the real testing method in detail. We are willing to practice and adjust a classifier to differentiate among pairs of pictures of the person "I" under examination and pairings of him or her with an imposter once we've picked the right collection of characteristics for every individual. A classifier's logic is that when a probing alleges to be "I," it would be paired with all of the components in the portfolio of I, and several test pairings as the gallery's size would be constructed. Such test pairings would be categorized to determine whether they reflect the similar individual I; due to feature choice, the categorization would be only dependent on the traits that matter to "I." We explore linear SVMs as a classifier once more, however, this time without the requirement for a coarse-to-fine approach because we just do one assessment per frame.

We operate every classifier on the testing set that we collected, which contains approximately 1700 probes, supposing an equivalent a prior probability of decent people and frauds (comparable to the assessment framework described in Messer et al., (2003): we utilize all positive data available for every classifier, and an adequate amount of negative are chosen randomly. Every testing picture T creates M testing pairings when it is passed through a particular classifier Ii. We calculate the proportion of testing pairs produced by every inquiry that were categorized as positives to assign every inquiry an exclusive output (Çeliktutan et al., 2013).

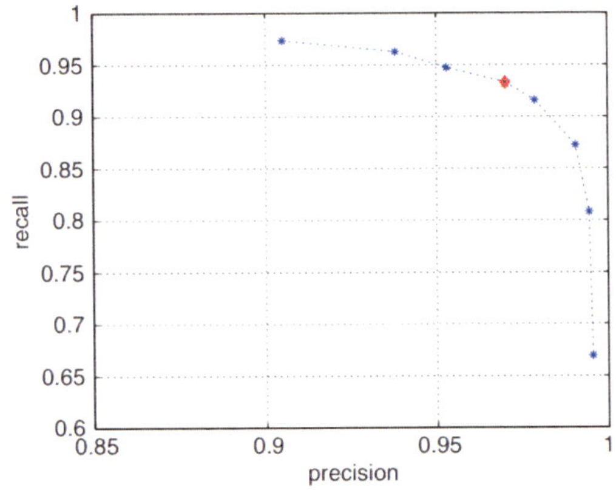

Figure 1.5. An example of a typical precision-recall behavior.

Source: https://link.springer.com/chapter/10.1007/978–3-642–35326–0_52.

Figure 1.5 illustrates the evolution of best accuracy by changing the percentage of positive scores q from 20% to 90%.

1.4.4. Quantitative Assessment of Multi-Task Learning

Multi-functional learning, like previously stated, can be utilized to learn groupings of similar features across several activities which might show distinct persons. We provide numeric simulations of the efficiency and durability of the features learned across an increasing range of jobs in this chapter.

We used the square loss function and used cross-validation to autonomously modify the regularization value. We take into account up to T = 200 regression activities. Any of such jobs' w_t variables were drawn from a 5D Gaussian distribution with zero mean and a covariance *Co*v = Diag (0.25, 0.36,0.49, 0.64,1) (Moghaddam, 1997). We keep adding 20 unrelated dimensions to such 5D w_t' s all of which are null. The test and training data were produced consistently from (0,1)d, with d ranging from 5 to 25. The y_{ti} outputs were calculated using the formula $y_{ti} = w_t, x_{ti}\rangle + \vartheta$, where ϑ is zero-mean Gaussian noise with a normal distribution of 0.1. As a result, the real characteristics $\langle u_i, x\rangle$ we wanted to learn were merely the input parameters in this example. On a 5 by 5 primary submatrix, the desired outcome is a feature matrix U that is near to the identity matrix (on 5 columns) and a matrix D that is generally proportionate to the covariance *Co*v utilized to create the activity variables (Mohan et al., 2001).

For testing and training, we created five and 20 samples per activity, accordingly. We applied our techniques with T = 10, 25, 100, and 200 activities to see how the quantity of jointly learned tasks affected test efficiency and, more significantly, the reliability of the features learned. We averaged the efficiency parameters over randomly chosen subsets of the 200 tasks for T = 10, 25, and 100, such that our predictions had identical changes. We also used conventional ridge regressions to assess every one of the 200 activities separately (Ojala et al., 2002).

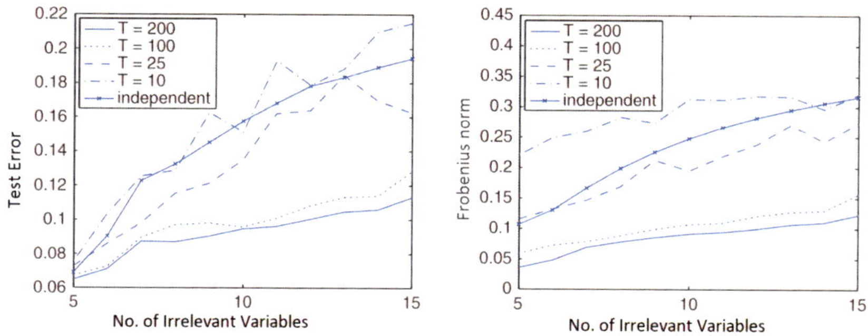

Figure 1.6. Left: Test error vs the number of external factors depending on various levels of activities (T). Right: Frobenius norm of the difference between learned and actualmatrices vs. the number of irrelevant variables (as the No. of tasks concurrently learnedchanges). This is a determination of the learned features' quality

Source: *https://link.springer.com/chapter/10.1007/978-1-84882-385-3_10.*

As the number of extraneous factors grows, the influence of the number of activities continuously learned on test efficiency and also the reliability of the features learned is shown in Figure 1.6. Firstly, as the left Figure illustrates, and in line with previous theoretical and empirical data – see, for example, (Baxter, 2000) – learning several activities simultaneously outperforms learning the tasks separately because the tasks are linked in this situation. Furthermore, as the number of tasks rises, performance improves. More significantly, as the number of extraneous factors increases, so does the improvement. For our objectives, the map on the right side of Figure 1.6 is the most significant. It indicates how muchthey taught features to differ from the individuals that were utilized to produce the data. We show the Frobenius norm of the discrepancy between the learned 5 by 5 principal submatrices of D and the real Cov matrix in further detail (standardize to have trace 1). We find that increasing the number of tasks contributes to more accurate estimations of the underlying characteristics, which is a major contribution of this section. Furthermore, the comparative (as the number of activities grows) quality of the features acquired improves with the number of extraneous variables, just as it does with test efficiency. Scheming the remaining of the learned U from the real one, and in this situation is the identity matrix, yielded similar findings (Zhao et al., 2003).

For six extraneous parameters, we additionally evaluated the influence of the regularization variable on the range of features acquired (as assessed using rank (D)). On the left side of Figure 1.7, we display the findings. The amount of characteristics acquired reduces with γ, as predicted. Lastly, the critical values of the components of matrix A obtained utilizing the variable determined using leave-one-out cross-validation are shown on the right side in Figure. 1.7. This is the resultant matrix across all 200 concurrently acquired activities and six extraneous parameters. This graph shows that our method generates a matrix *A* with the anticipated design: just five key characteristics (Nomura et al., 2021).

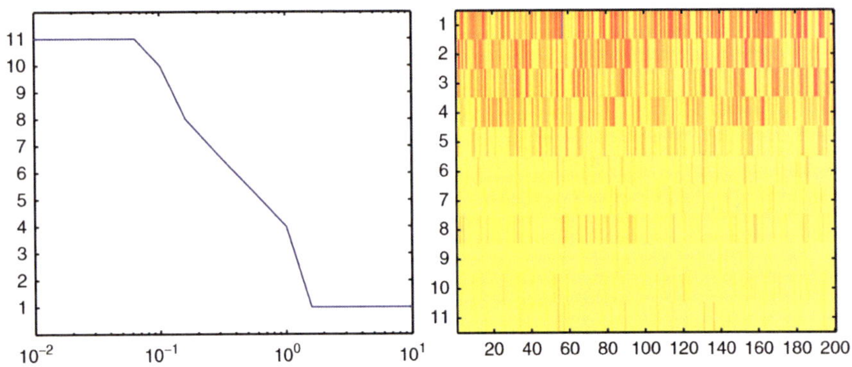

Figure 1.7. Linear synthetic data. Left: The range of features acquired vs the regularization variable γ (notice transcript for explanation). Right: matrix A, indicating the importance of the learned features – the first five rows correspond to the true features. The color scale ranges from yellow (lower values) to purple (higher values).

Source: *https://link.springer.com/chapter/10.1007/978-3-642-35326-0_52.*

The actual values (diagonal entries of *Cov* scaled to have tracing 1) are 0.36, 0.23, 0.18, 0.13, and 0.09, correspondingly, whereas the (standardize) 2nd norms of the adjacent rows are 0.31, 0.21, 0.12, 0.10, and 0.09, correspondingly (Tyagi et al., 2018).

REFERENCES

1. Abdelwhab, A., & Viriri, S. (2018). A survey on soft biometrics for human identification. *Machine Learning and Biometrics, 37, 1–17.*
2. Aggarwal, G., Chowdhury, A. R., & Chellappa, R. (2004). A system identification approach for video-based face recognition. In *Proceedings of the 17th International Conference on Pattern Recognition,* 4, 175–178.
3. Ahonen, T., Hadid, A., & Pietikainen, M. (2006). Face description with local binary patterns: Application to face recognition. *IEEE Transactions on Pattern Analysis and Machine Intelligence*, *28*(12), 2037–2041.
4. Alom, M. M., Begum, N., Ferdousi, S., Begum, S., & Ali, T. (2009). Power spectral analysis of heart rate variability in adolescent male athletes. *Journal of Bangladesh Society of Physiologist*, *4*(2), 26–33.
5. Anaya-Esparza, L. M., González-Aguilar, G. A., Domínguez-Ávila, J. A., Olmos-Cornejo, J. E., Pérez-Larios, A., & Montalvo-González, E. (2018). Effects of minimal processing technologies on jackfruit (Artocarpus heterophyllus Lam.) quality parameters. *Food and Bioprocess Technology*, *11*(9), 1761–1774.
6. Ando, R. K., Zhang, T., & Bartlett, P. (2005). A framework for learning predictive structures from multiple tasks and unlabeled data. *Journal of Machine Learning Research*, *6*(11), 1–19.
7. Arca, S., Campadelli, P., & Lanzarotti, R. (2006). A face recognition system based on automatically determined facial fiducial points. *Pattern Recognition*, *39*(3), 432–443.
8. Argyriou, A., Evgeniou, T., & Pontil, M. (2008). Convex multi-task feature learning. *Machine Learning*, *73*(3), 243–272.
9. Baxter, J. (2000). A model of inductive bias learning. *Journal of Artificial Intelligence Research*, *12*, 149–198.
10. Beam, A. L., & Kohane, I. S. (2018). Big data and machine learning in health care. *JAMA*, *319*(13), 1317–1318.
11. Belhumeur, P. N., Hespanha, J. P., & Kriegman, D. J. (1997). Eigenfaces vs. fisherfaces: Recognition using class specific linear projection. *IEEE Transactions on Pattern Analysis and Machine Intelligence*, *19*(7), 711–720.
12. Bicego, M., Lagorio, A., Grosso, E., & Tistarelli, M. (2006). On the use of SIFT features for face authentication. In *2006 Conference on Computer Vision and Pattern Recognition Workshop, 2(1),* 35–35.

13. Biju, S., Fuentes, S., Gonzalez Viejo, C., Torrico, D. D., Inayat, S., & Gupta, D. (2021). Silicon supplementation improves the nutritional and sensory characteristics of lentil seeds obtained from drought□ stressed plants. *Journal of the Science of Food and Agriculture*, *101*(4), 1454–1466.
14. Bisschoff, C. A., Coetzee, B., & Esco, M. R. (2018). Heart rate variability and recovery as predictors of elite, African, male badminton players' performance levels. *International Journal of Performance Analysis in Sport*, *18*(1), 1–16.
15. Brunelli, R., & Poggio, T. (1993). Face recognition: Features versus templates. *IEEE Transactions on Pattern Analysis and Machine Intelligence*, *15*(10), 1042–1052.
16. Çeliktutan, O., Ulukaya, S., & Sankur, B. (2013). A comparative study of face landmarking techniques. *EURASIP Journal on Image and Video Processing*, *2013*(1), 1–27.
17. Chen, S. S., Donoho, D. L., & Saunders, M. A. (2001). Atomic decomposition by basis pursuit. *SIAM Review*, *43*(1), 129–159.
18. Cherkassky, V., & Ma, Y. (2009). Another look at statistical learning theory and regularization. *Neural Networks*, *22*(7), 958–969.
19. Daubechies, I., Defrise, M., & De Mol, C. (2004). An iterative thresholding algorithm for linear inverse problems with a sparsity constraint. *Communications on Pure and Applied Mathematics: A Journal Issued by the Courant Institute of Mathematical Sciences*, *57*(11), 1413–1457.
20. Destrero, A., De Mol, C., Odone, F., & Verri, A. (2007). A regularized approach to feature selection for face detection. In *Asian Conference on Computer Vision*, 3(1), 881–890.
21. Destrero, A., De Mol, C., Odone, F., & Verri, A. (2009). A regularized framework for feature selection in face detection and authentication. *International Journal of Computer Vision*, *83*(2), 164–177.
22. Destrero, A., Mosci, S., De Mol, C., Verri, A., & Odone, F. (2009). Feature selection for high-dimensional data. *Computational Management Science*, *6*(1), 25–40.
23. Donoho, D. L. (2000). High-dimensional data analysis: The curses and blessings of dimensionality. *AMS Math Challenges Lecture*, *1*(2000), 32.

24. Edwards, G. J., Taylor, C. J., & Cootes, T. F. (1999). Improving identification performance by integrating evidence from sequences. In *Proceedings. 1999 IEEE Computer Society Conference on Computer Vision and Pattern Recognition,* 1, 486–491.
25. Etemad, K., & Chellappa, R. (1997). Discriminant analysis for recognition of human face images. *Josa A*, *14*(8), 1724–1733.
26. Evgeniou, A., & Pontil, M. (2007). Multi-task feature learning. *Advances in Neural Information Processing Systems*, *19*, 41.
27. Farahmand, E., Razavi, S. H., & Mohtasebi, S. S. (2021). Investigating effective variables to produce desirable aroma in sourdough using e□ nose and sensory panel. *Journal of Food Processing and Preservation*, *45*(2), 15157.
28. Fardin, P., Barla, A., Mosci, S., Rosasco, L., Verri, A., & Varesio, L. (2009). The L 1-L 2 regularization framework unmasks the hypoxia signature hidden in the transcriptome of a set of heterogeneous neuroblastoma cell lines. *BMC Genomics*, *10*(1), 1–16.
29. Fazel, M., Hindi, H., & Boyd, S. P. (2001). A rank minimization heuristic with application to minimum order system approximation. In *Proceedings of the 2001 American Control Conference,* 6, 4734–4739.
30. Friedman, J., Hastie, T., & Tibshirani, R. (2000). Additive logistic regression: a statistical view of boosting (with discussion and a rejoinder by the authors). *The Annals of Statistics*, *28*(2), 337–407.
31. Friedman, J., Hastie, T., & Tibshirani, R. (2010). Regularization paths for generalized linear models via coordinate descent. *Journal of Statistical Software*, *33*(1), 1.
32. Fuentes, S., Wong, Y. Y., & Gonzalez Viejo, C. (2020). Non-invasive biometrics and machine learning modeling to obtain sensory and emotional responses from panelists during entomophagy. *Foods*, *9*(7), 903.
33. Gonzalez Viejo, C., Villarreal-Lara, R., Torrico, D. D., Rodríguez–velazco, Y. G., Escobedo-Avellaneda, Z., Ramos-Parra, P. A., ... & Fuentes, S. (2020). Beer and consumer response using biometrics: Associations assessment of beer compounds and elicited emotions. *Foods*, *9*(6), 821.
34. Granata, F. (2019). Evapotranspiration evaluation models based on machine learning algorithms—A comparative study. *Agricultural Water Management*, *217*, 303–315.

35. Guyon, I., & Elisseeff, A. (2003). An introduction to variable and feature selection. *Journal of Machine Learning Research*, *3*(Mar), 1157–1182.
36. He, X., Yan, S., Hu, Y., Niyogi, P., & Zhang, H. J. (2005). Face recognition using *Laplacian faces. IEEE Transactions on Pattern Analysis and Machine Intelligence*, *27*(3), 328–340.
37. Heiselet, B., Serre, T., Pontil, M., & Poggio, T. (2001). Component-based face detection. In *Proceedings of the 2001 IEEE Computer Society Conference on Computer Vision and Pattern Recognition,* 1, 1–4.
38. Hengl, T., Mendes de Jesus, J., Heuvelink, G. B., Ruiperez Gonzalez, M., Kilibarda, M., Blagotić, A., ... & Kempen, B. (2017). SoilGrids250m: Global gridded soil information based on machine learning. *PLoS one*, *12*(2), 0169748.
39. Hernández-Santos, B., Santiago-Adame, R., Navarro-Cortéz, R. O., Gómez-Aldapa, C. A., Castro-Rosas, J., Martínez-Sánchez, C. E., ... & Rodríguez-Miranda, J. (2015). Physical properties of ebony seed (*Pithecellobium flexicaule*) and functional properties of whole and defatted ebony seed meal. *Journal of Food Science and Technology*, *52*(7), 4483–4490.
40. Jain, A. K., Ross, A., & Prabhakar, S. (2004). An introduction to biometric recognition. *IEEE Transactions on Circuits and Systems for Video Technology*, *14*(1), 4–20.
41. Juárez-Barrientos, J. M., Hernández-Santos, B., Herman-Lara, E., Martínez-Sánchez, C. E., Torruco-Uco, J. G., de Jesus Ramírez-Rivera, E., ... & Rodríguez-Miranda, J. (2017). Effects of boiling on the functional, thermal and compositional properties of the Mexican jackfruit (Artocarpus heterophyllus) seed Jackfruit seed meal properties. *Emirates Journal of Food and Agriculture*, 1–9, 4–19.
42. Kampmann, M., & Zhang, L. (1998). Estimation of eye, eyebrow and nose features in videophone sequences. In *International Workshop on Very Low Bitrate Video Coding, 98*, 101–104.
43. Leger, S., Zwanenburg, A., Pilz, K., Lohaus, F., Linge, A., Zöphel, K., ... & Richter, C. (2017). A comparative study of machine learning methods for time-to-event survival data for radiomics risk modelling. *Scientific Reports*, *7*(1), 1–11.

44. Lemley, J., Bazrafkan, S., & Corcoran, P. (2017). Deep Learning for Consumer Devices and Services: Pushing the limits for machine learning, artificial intelligence, and computer vision. *IEEE Consumer Electronics Magazine*, *6*(2), 48–56.
45. Li, B., & Chellappa, R. (2001). Face verification through tracking facial features. *JOSA A*, *18*(12), 2969–2981.
46. Li, S. Z., & Zhang, Z. (2004). Floatboost learning and statistical face detection. *IEEE Transactions on Pattern Analysis and Machine Intelligence*, *26*(9), 1112–1123.
47. Lichters, M., Möslein, R., Sarstedt, M., & Scharf, A. (2021). Segmenting consumers based on sensory acceptance tests in sensory labs, immersive environments, and natural consumption settings. *Food Quality and Preference*, *89*, 104138.
48. Liu, X., Chen, T., & Kumar, B. V. (2002, May). On modeling variations for face authentication. In *Proceedings of Fifth IEEE International Conference on Automatic Face Gesture Recognition*, *2*(1), 384–389.
49. Madi-Jebara, S. N., Sleilaty, G. S., Achouh, P. E., Yazigi, A. G., Haddad, F. A., Hayek, G. M., ... & Jebara, V. A. (2004). Postoperative intravenous iron used alone or in combination with low-dose erythropoietin is not effective for correction of anemia after cardiac surgery. *Journal of Cardiothoracic and Vascular Anesthesia*, *18*(1), 59–63.
50. Manresa-Yee, C., Varona, J., Perales, F. J., & Salinas, I. (2014). Design recommendations for camera-based head-controlled interfaces that replace the mouse for motion-impaired users. *Universal Access in the Information Society*, *13*(4), 471–482.
51. Meng, X., Bradley, J., Yavuz, B., Sparks, E., Venkataraman, S., Liu, D., ... & Talwalkar, A. (2016). Mllib: Machine learning in apache spark. *The Journal of Machine Learning Research*, *17*(1), 1235–1241.
52. Messer, K., Kittler, J., Sadeghi, M., Marcel, S., Marcel, C., Bengio, S., ... & Mavity, N. (2003). Face verification competition on the XM2VTS database. In *International Conference on Audio-and Video-Based Biometric Person Authentication*, 4(12), 964–974.
53. Moghaddam, B., &, A. (1997). Probabilistic visual learning for object representation. *IEEE Transactions on pattern analysis and machine intelligence*, *19*(7), 696–710.

54. Moghaddam, B., & Pentland, A. P. (1994, October). Face recognition using view-based and modular eigenspaces. In *Automatic Systems for the Identification and Inspection of Humans*, *2277*, 12–21.
55. Mohamed, A. E. (2017). Comparative study of four supervised machine learning techniques for classification. *International Journal of Applied*, *7*(2), 1–10.
56. Mohan, A., Papageorgiou, C., & Poggio, T. (2001). Example-based object detection in images by components. *IEEE transactions on pattern analysis and machine intelligence*, *23*(4), 349–361.
57. Nomura, M., Osada, E., Tokita, T., Iwamoto, T., & Manome, Y. (2021). Measurement and differentiation of banana juice scent using an electronic nose FF-2A. *PeerJ*, *9*, 10638.
58. Ojala, T., Pietikainen, M., & Maenpaa, T. (2002). Multiresolution gray-scale and rotation invariant texture classification with local binary patterns. *IEEE Transactions on Pattern Analysis and Machine Intelligence*, *24*(7), 971–987.
59. Ouamane, A., Messaoud, B., Guessoum, A., Hadid, A., & Cheriet, M. (2014). Multi scale multi descriptor local binary features and exponential discriminant analysis for robust face authentication. In *2014 IEEE International Conference on Image Processing, 3(3),* 313–317.
60. Palaniappan, R., & Mandic, D. P. (2007). Biometrics from brain electrical activity: A machine learning approach. *IEEE Transactions on Pattern Analysis and Machine Intelligence*, *29*(4), 738–742.
61. Perrett, D. I., Mistlin, A. J., Potter, D. D., Smith, P. A. J., Head, A. S., Chitty, A. J., ... & Jeeves, M. A. J. (1986). Functional organization of visual neurones processing face identity. In *Aspects of Face Processing*, *3*(1), 187–198.
62. Pires, M. A., Rodrigues, I., Barros, J. C., & Trindade, M. A. (2021). Kelly's repertory grid method applied to develop sensory terms for consumer characterization (check-all-that-apply) of omega-3 enriched bologna sausages with reduced sodium content. *European Food Research and Technology*, *247*(1), 285–293.
63. Pontil, M., & Verri, A. (1998). Support vector machines for 3D object recognition. *IEEE Transactions on Pattern Analysis and Machine Intelligence*, *20*(6), 637–646.

64. Pontile, M., Evgeniou, T., & Argyriou, A. (2007). Convex multi-task feature learning. *Journal of Machine Learning*, *10*, 243–272.
65. Poostchi, M., Silamut, K., Maude, R. J., Jaeger, S., & Thoma, G. (2018). Image analysis and machine learning for detecting malaria. *Translational Research*, *194*, 36–55.
66. Ranftl, A., Alonso-Fernandez, F., Karlsson, S., & Bigun, J. (2017). Real-time AdaBoost cascade face tracker based on likelihood map and optical flow. *IET Biometrics*, *6*(6), 468–477.
67. Roth, D., Yang, M. H., & Ahuja, N. (2000). A snow-based face detector. In *Neural Information Processing*, *12*, 3–19.
68. Rowley, H. A., Baluja, S., & Kanade, T. (1998). Neural network-based face detection. *IEEE Transactions on pattern analysis and machine intelligence*, *20*(1), 23–38.
69. Schapire, R., & Freund, Y. (1995). A decision-theoretic generalization of on-line learning and an application to boosting. *Second European Conference on Computational Learning Theory*, *3*(1), 23–37.
70. Schneiderman, H., & Kanade, T. (2000, June). A statistical method for 3D object detection applied to faces and cars. *Proceedings IEEE Conference on Computer Vision and Pattern Recognition, 1*, 746–751.
71. Srebro, N., Rennie, J., & Jaakkola, T. (2005). Maximum-margin matrix factorization. *Advances in Neural Information Processing Systems, 2*(1), 1–14.
72. Sundararajan, K., & Woodard, D. L. (2018). Deep learning for biometrics: A survey. *ACM Computing Surveys (CSUR)*, *51*(3), 1–34.
73. Sung, K. K., & Poggio, T. (1998). Example-based learning for view-based human face detection. *IEEE Transactions on Pattern Analysis and Machine Intelligence*, *20*(1), 39–51.
74. Tibshirani, R. (1996). Regression shrinkage and selection via the lasso. *Journal of the Royal Statistical Society: Series B (Methodological)*, *58*(1), 267–288.
75. Ting, H. Y., Sim, K. S., & Abas, F. S. (2015). Kinect-based badminton movement recognition and analysis system. *International Journal of Computer Science in Sport*, *14*(2), 25–41.
76. Tiwari, D., & Tyagi, V. (2016). A novel scheme based on local binary pattern for dynamic texture recognition. *Computer Vision and Image Understanding*, *150*, 58–65.

77. Torralba, A., Murphy, K. P., & Freeman, W. T. (2004, June). Sharing features: efficient boosting procedures for multiclass object detection. In *Proceedings of the 2004 IEEE Computer Society Conference on Computer Vision and Pattern Recognition,* 2, 1–17.
78. Turk, M., & Pentland, A. (1991). Eigenfaces for recognition. *Journal of Cognitive Neuroscience*, *3*(1), 71–86.
79. Tyagi, R., Tiwari, A., Gupta, A. K., & Gupta, S. (2018). Transcriptome wide identification and characterization of Starch Synthase enzyme in finger millet. *Bioinformation*, *14*(7), 393.
80. Tzavella, L., Maizey, L., Lawrence, A. D., & Chambers, C. D. (2020). The affective priming paradigm as an indirect measure of food attitudes and related choice behavior. *Psychonomic bulletin & review*, *27*(6), 1397–1415.
81. Ullman, S., Vidal-Naquet, M., & Sali, E. (2002). Visual features of intermediate complexity and their use in classification. *Nature Neuroscience*, *5*(7), 682–687.
82. Viola, P., & Jones, M. J. (2004). Robust real-time face detection. *International Journal of Computer Vision*, *57*(2), 137–154.
83. Wahab, L., & Jiang, H. (2019). A comparative study on machine learning based algorithms for prediction of motorcycle crash severity. *PLoS One*, *14*(4), e0214966.
84. Weston, J., Elisseeff, A., Schölkopf, B., & Tipping, M. (2003). Use of the zero norm with linear models and kernel methods. *The Journal of Machine Learning Research*, *3*, 1439–1461.
85. Wiskott, L., Krüger, N., Kuiger, N., & Von Der Malsburg, C. (1997). Face recognition by elastic bunch graph matching. *IEEE Transactions on Pattern Analysis and Machine Intelligence*, *19*(7), 775–779.
86. Yang, M. H., Kriegman, D. J., & Ahuja, N. (2002). Detecting faces in images: A survey. *IEEE Transactions on Pattern Analysis and Machine Intelligence*, *24*(1), 34–58.
87. Yiiong, S. P., Ting, H. Y., Tan, D. Y. W., & Chia, R. (2019, April). Investigation of Relation between Sport's Motion and Heart Rate Variability (HRV) Based on Biometric Parameters. In *IOP Conference Series: Materials Science and Engineering*, 495(1), 12015.
88. Yu, B., & Xu, Z. B. (2008). A comparative study for content-based dynamic spam classification using four machine learning algorithms. *Knowledge-Based Systems*, *21*(4), 355–362.

89. Yuan, S., & Chu, F. (2007). Fault diagnosis based on support vector machines with parameter optimisation by artificial immunisation algorithm. *Mechanical Systems and Signal Processing*, *21*(3), 1318–1330.
90. Zhang, P., Wu, H. N., Chen, R. P., & Chan, T. H. (2020). Hybrid meta-heuristic and machine learning algorithms for tunneling-induced settlement prediction: a comparative study. *Tunnelling and Underground Space Technology*, *99*, 103383.
91. Zhao, W., Chellappa, R., Phillips, P. J., & Rosenfeld, A. (2003). Face recognition: a literature survey. *ACM Computing Surveys (CSUR)*, *35*(4), 399–458.

Chapter 2

Biometric Detection Recognition Techniques

CONTENTS

2.1. INTRODUCTION

Currently, automation is present in every aspect of our lives. However, automated verification systems based on traditional passwords can be hacked easily when the password is revealed to an unapproved user. Several biometric methods have been offered for personal identification systems. Biometric-centered personal authentication systems utilize behavioral or physiological individualities of the person for the aim of recognition (Masmoudi et al., 2010). Physiological individualities comprise fingerprint, hand geometry, face, and behavioral individualities are handwriting, speech. Throughout the last few years, an outstanding advancement in the technology of biometric recognition has been achieved because of the increasing necessity of highly trustworthy personal authentication and recognition in various commercial and government applications. A relative study of several biometric identifications techniques has been suggested in this chapter for authentication and also to minimize fraud.

Biometric identification is described as an automated identification for individuals based on their biological features, such as fingerprints, face, voice, and gait (Bhattacharyya et al., 2009). The earliest scientific publication on biometric identification as a result of fingerprint matching was published in 1963.

Normally, the biometric individualities are intrinsic to the individual, there exists an everlasting association between the individual and his/her individualities. Therefore, personal recognition can be done with the help of one or more biometric individualities. The logical procedure can be accomplished through scrutiny operations, where identification is needed to attain their identity to store it and then utilize this stored data to recognize the individual as required. Furthermore, in present recognition systems, the restrictions that exist in the utilization of passwords are removed by utilizing these biometric recognition systems, where a system needs the individual to give his/her features that can't be presented by others (Joshi et al., 2015).

Biometrics identification systems are in incessant development and are an outcome of usability of execution from day to day. Therefore, biometric identification evaluation is a continuous process that has become more interesting as an outcome of its serviceability (Truta et al., 2012).

Biometrics is the technique for identifying human traits, both behavioral and physiological. Confirmation of individuals centered on biometric authentication is becoming progressively popular in several applications such as banking, financial transactions, and aviation. The diagram of the biometric identification system is displayed in Figure 2.1. The key functioning of this system includes two sections; enrolment and test. Throughout the process of enrolment, the template gets saved in a database and during the test procedure, the data of individuals are matched with the attained templates. The matching program assesses the template with input, approximating the distance between the two utilizing the appropriate algorithm. This is well-thought-out as an output for a particular purpose (Depren et al., 2005).

The 1st block, called the sensor behaves as an interface amongst the system and the real world and attains essential data. A picture acquisition system based on vision is a suitable choice for it. The second block carries out preprocessing, eliminates artifacts from the sensor, and improves input pictures. The necessary features are extracted in the next block. This is a serious stage as the right characteristics are needed to be dug out in an optimum way. Image characteristics of the vector of numbers with particular properties are utilized for the creation of the template. The template is a combinational set of associated characteristics obtained from the source. The essentials of biometric measurements that aren't needed for comparing algorithms are expelled in templates for decreasing the file size and for shielding the individuality of the claimer (Farnaaz & Jabbar, 2016).

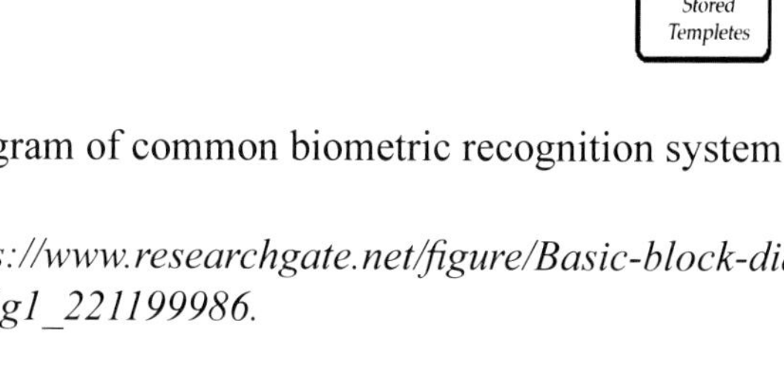

Figure 2.1. Diagram of common biometric recognition system.

Source: *https://www.researchgate.net/figure/Basic-block-diagram-of-a-biometric-system_fig1_221199986.*

2.2. PHYSIOLOGICAL QUALITIES

Physiological qualities are associated with the physical structure of an individual. The multiplicity of physical properties associated with the human body are used for verification purposes, such as fingerprints, hand and facial geometry, retina, and vascular patterns. Computer vision is a popular tool to detect these properties (Jabez & Muthukumar, 2015).

2.2.1. Finger Prints

The detection of fingerprints includes either recognition of minutiae specifically ridge ending, dot, bifurcation, or an island (Figure 2.2), or pattern matching centered on a vision for identification.

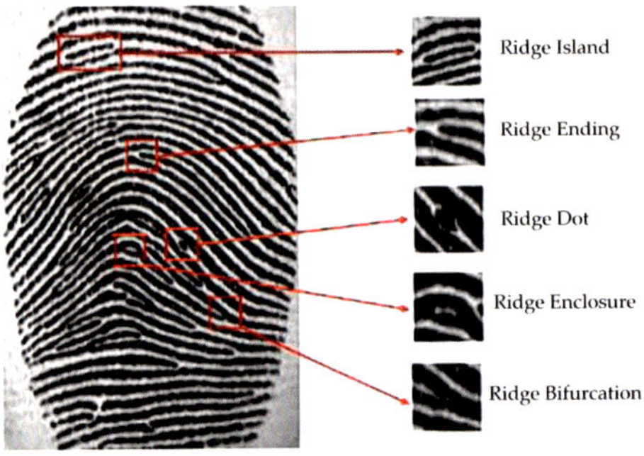

Figure 2.2. Minutiae configuration of the fingerprint.

Source: *https://www.bayometric.com/minutiae-based-extraction-fingerprint-recognition/.*

Techniques centered on minutiae are dependent on the recognition of minutiae direction and location for detection while pattern matching matches two images to find out the similarities.

Techniques utilized for fingerprint detection are centered on optical, thermal, ultrasonic, or capacitance principles (Bhattacharyya et al., 2009).

Optical techniques depend on taking the digital image created by light reflection from the ends where edges touch the touch surface of the sensor. Optical fingerprint detectors include the light sensor, light source, touch surface where the finger is located, and the capture device which generally can be charge-coupled with the CMOS camera, or a simple webcam.

The capacitive fingerprint sensors comprise a group of capacitive plates on the silicon chip. One capacitor plate is created by the finger while the other plate comprises a small metallization area on the chip. When a finger is placed on the surface of a chip, the edges of a finger are near to the adjacent pixels and thus possess high capacitance. Comparatively, the valleys are distant from the adjacent pixels and therefore possess lower capacitance.

An ultrasonic technique utilized high-frequency sound waves to monitor the surfaces of a finger, the individual places his/her finger on the glass piece, and an ultrasonic sensor then moves and detects the entire fingerprint. The tip of a finger is applied to the window of an ultrasonic head. The ultrasonic technique of attaining fingerprint is centered on sending the ultrasonic signals in the direction of the finger and perceiving the echo (Sondhiya et al., 2013).

A thermal method is primarily beneficial for the recognition of hidden fingerprints. The research exhibited that sebaceous rich and eccrine rich latent print imprint on paper might be made noticeable by applying heat between 220–300°C. When the heated paper substrate is perceived under a light in 505 nanometers light range the latent impression can be made observable.

The method of fingerprint detection provides the following advantages (Gul & Hussain, 2011):

i. It offers high uniqueness. All human beings have different fingerprints. Twins, who possess similar patterns of DNA, have distinct fingerprints. Thus fingerprints are the strong verification mechanism.
ii. The patterns of fingerprint specifically ridges are created in the womb and stay invariant through lifespan except in a situation of very deep injury.
iii. Excluding an amputee, human beings have readable fingerprints so can be verified easily without the necessity of carrying the token.
iv. It is an extensively accepted method.

The limitation comprises duplication of fingerprints, exposure to distortion, and noise due to twists and dirt. Also, few people don't find it suitable to place their fingers on the exact location which has been touched by several other people.

2.2.2. Facial Recognition

It can be considered as the computer application for confirming an individual from the digital image or the video capture. This method depends on the mapping of particular facial characteristics such as the nose width, distance between the eyes, and length of jawlines. These are termed nodal points and are generally measured by making numerical code which is the face print for a person. This face print signifies the face in a database. Both two-dimensional and three-dimensional techniques are utilized for automated facial identification systems (Zarrabi & Zarrabi, 2012).

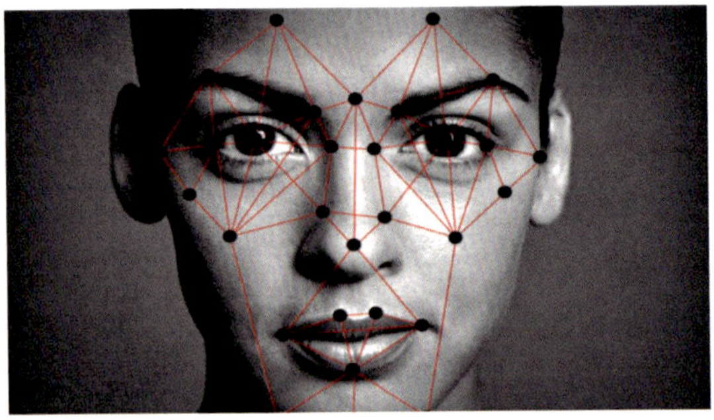

Figure 2.3. Face's nodal points.

Source: https://hhsbearpress.com/2018/05/14/all-about-facial-recognition/.

2.2.2.1. Two-Dimensional Facial Recognition

In this technique, the two-dimensional picture of the face is matched with the one saved in the database. The picture is taken with the help of a camera and is signified as the vector of intensities. The vector is then estimated as the sum of basis vectors calculated by principal component examination from the database of face pictures. The principal components signify the usual variations observed between faces and offer a brief encapsulation of the emergence of the sample face picture, and the basis for its contrast with other face pictures (Masmoudi et al., 2010). From the processed picture, characteristics are taken out. This extracted characteristic is known as face template or facial code. The face template is then matched with the database and resemblances are determined. For precision, the image must be taken with a face looking towards the camera. Issues can exist with little discrepancy of facial expression or light from the pictures saved in the database. As matched to the three-dimensional facial recognition method it requires lesser space of storage for recognition templates. 2D pictures comprise inadequate information and are sensitive to expressions, orientation, and illumination (Chihaoui et al., 2016).

2.2.2.2. Three-Dimensional Facial Recognition

This technique utilizes the real-time capturing of the facial picture and utilizes the characteristics of the face such as curves of the nose, chin, and eye socket where firm bone and tissue are most obvious. As this method utilizes the depth and axis of measurement that isn't disturbed by lighting, thus it can be used to identify the subject at distinct view angles (Figure 2.4).

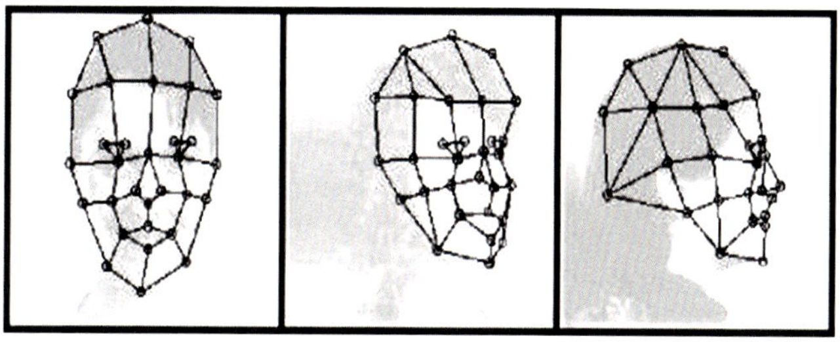

Figure 2.4. Characteristics for 3D identification.

Source: *https://www.ijareeie.com/upload/october/10_Biometric%20Recognition%20Techniques.pdf.*

Correctness of facial demonstration is high with the capability to take and save more information. The utilization of three-dimensional data offers much better management of orientation and illumination associated variations (Rath & Rautaray, 2014).

The cost of computation is high as a huge amount of data needs to be processed.

2.2.3. Iris Identification

The iris of an eye is a colored annular area that encloses the pupil (Figure 2.5). The patterns of the iris are exclusive. No 2 irises are similar, even the left and right eye of the same individual.

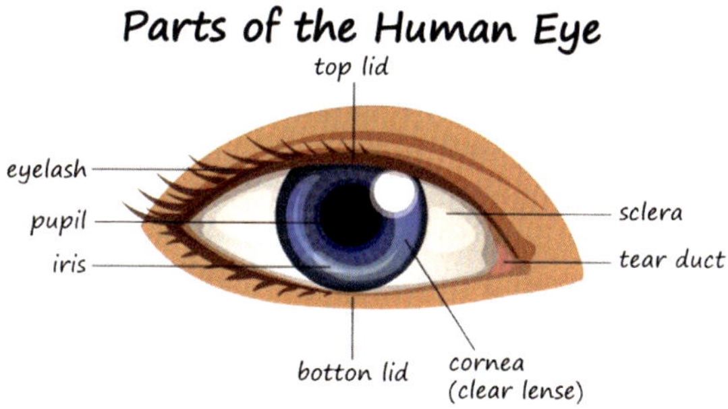

Figure 2.5. Representation of Iris.

Source: *https://www.freepik.com/free-photos–vectors/eye-iris.*

The technique of iris detection might involve the acquisition of image, localization, breakdown, and matching. Iris image with high resolution is needed for authentication, so the camera with such abilities and high transfer rate of frame or video making device can be utilized. After attaining the iris image, the step of localization is carried out. By this step, the iris portion of an image is observed. This can be estimated by two circles, the first one is the outer boundary of the iris (sclera), and the other one internal to the first is the inner boundary of the iris (pupil) (Joshi et al., 2015). Localization of iris is trailed by segmentation which can generally be defined as the procedure of splitting the input iris picture into various components. After localization and segmentation of the iris, the last stage is pattern matching of an iris image with the saved templates in the database.

This method of person verification is highly precise because iris patterns don't vary over time. Attaining an iris picture makes this technique problematic for implantation with simplicity as it needs proper orientation and positioning.

2.2.4. Hand Geometry and Fist Print

Recognition systems centered on hand geometry use the geometric characteristic of the hand such as width and length of fingers, palm diameter, and perimeter. The techniques based on vision are used for authentication (Shaheed et al., 2018). It comprises the acquisition of the image, extraction of feature, matching of template, as in the situation of other biometric techniques. The acquisition of images for hand biometrics might contact kind and the guided one, which needs the flat platform to keep the hand and hooks to assists the placement of the user's hand, or the non-contact techniques. Controlled and contact-centered systems needing the flat platform and pins to confine the freedom of hand. Uncontrolled and contact-centered peg-free scenarios, even though still needing the platform to keep the hand. Uncontrolled and contact-free. Contact-less situations where neither platform nor pegs are needed for acquiring of hand image (Santos et al., 2011).

Prints of palm are sequences of dark lines signifying peaking shares of friction corrugated skin. Either techniques based on minutiae which are dependent on the direction, location, and alignment of minutiae points, or matching based on a ridge that uses ridge patterns characteristics like sweat pores, geometrical traits, and spatial features can be utilized (Chihaoui et al., 2016).

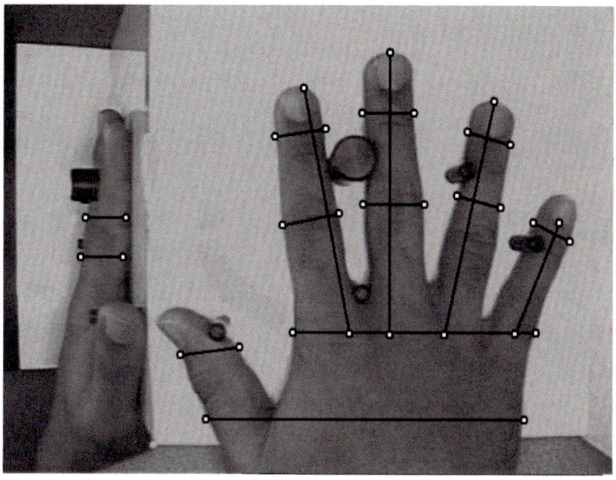

Figure 2.6. Geometry of hand.

Source: *https://www.m2sys.com/blog/guest-blog-posts/about-hand-geometry-identification/.*

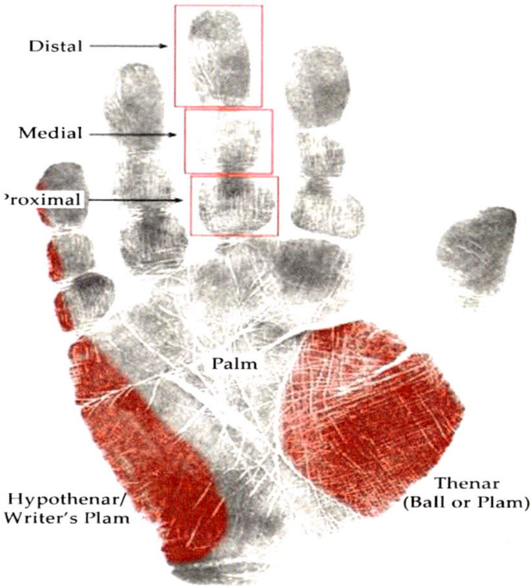

Figure 2.7. Print of Palm for biometric recognition.

Source: *https://link.springer.com/chapter/10.1007/978–3-319–10365–5_4.*

2.2.5. Retina Identification

Retina to an eye is just like the film is to a camera. It has billions of photoreceptors that convert light rays into electrical impulses.

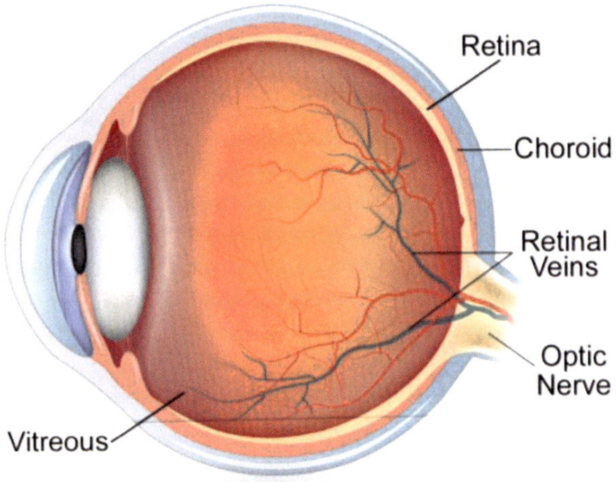

Figure 2.8. Human Retina

Source: *https://www.vmrinstitute.com/what-is-the-retina/*

Recognition technique based on retina utilizes a pattern of a blood vessel in a retina which is exclusive for the eye. As the retina is on the backside of an eye, thus it isn't visible directly and that's why is stable over the lifespan of an individual. Further due to this reason the IR light source is essential to brighten the retina (Guerra et al., 2011). The benefit of IR light is that blood vessels in the retina absorb it faster as compared to the remaining portions of eye tissue. The light reflected is consequently taken by the scanning device in order to process it further.

The procedure of retina scanning includes the acquisition of image and processing, pattern matching, and illustration of it in the form of a template. The template size must be very small and is the smallest amongst the other biometrics techniques.

The scans of the retina necessitate that the individual removes his/her glasses, keeps their eye nearby the scanner, looks at a particular point, and stays still, and concentrates on a particular location for nearly 10–16 seconds in order to complete the scan (Róka et al., 2007).

2.2.6. Pattern Identification of Hand Vein

Vein identification systems primarily emphasize vascular patterns in the hands of users. Matched to some other biometric recognition systems, the veins of users are located within the human body so hard it is to replicate, therefore vein verification technology provides a high level of precision.

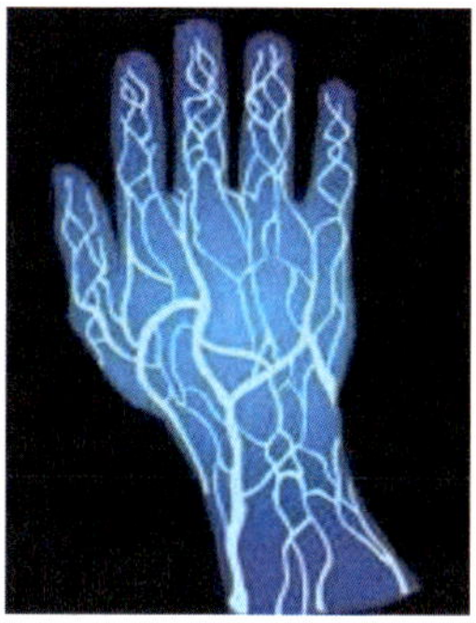

Figure 2.9. Vascular pattern of human hand.

Source: http://www.dalle–vedove.com/modalities/vascular/index.php.

NIR (near-infrared) rays from the bank of light emanating diodes enter the skin of the hand and yield an image triggered by the absorbance of the blood vessels. This picture is digitized in order to make distinct templates, which creates the database of the biometric device. Several features utilized for templates are the branching points of the vessel, veins thickness, and the branching angles. The devices for vascular imaging can be made in either contacting kind or function in a non-contact fashion. The non-contacting technique provides the advantage in that it isn't essential for the person to touch the sensor to give biometric data. This is beneficial in applications where hygiene is to be preserved, like access to a medical operating room or where individuals are sensitive regarding touching the biometric sensing device (Kaur & Madaan, 2015).

2.3. BEHAVIORAL QUALITIES

2.3.1. Dynamic Signature Identification

Day-to-day behaviors can also serve as an identification tool. For the real-time signature identification, customers put their signature on the digital tablet which is usually connected to the personal computer for further processing and authentication (Walker, 2012). The patterns of behavior inherent to the procedure of signing comprise the included timing, pressure applied while writing, and speed. Even though it is relatively simple to replicate the visual look of the signature, it is hard to replicate behavioral features. The dynamic information utilized for purpose of identification normally comprises spatial coordinates, azimuth, pressure, inclination, acceleration, and velocity.

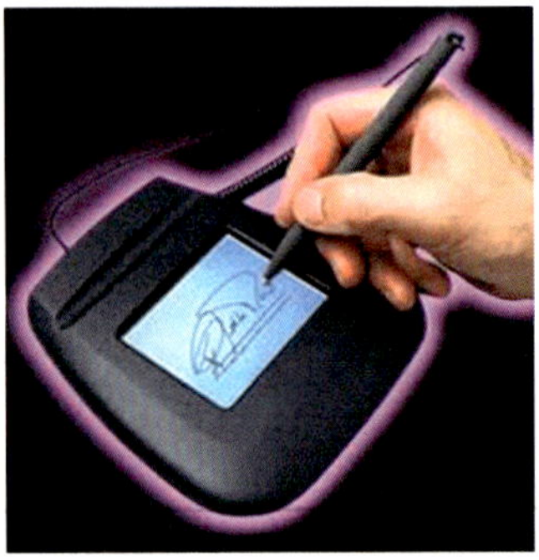

Figure 2.10. Information withdrawal for signature identification.

Source: https://www.biometricupdate.com/201206/explainer-dynamic-signature.

The most substantial advantage of Signature Identification is that it is very resistant to frauds, as it is very easy to forge the signature, however it is very hard to mimic the patterns of behavior inherent to the procedure of signing. Conversely, Signature Identification is susceptible to higher rates of error, chiefly when the behavioral features of signatures are equally unreliable (Fang et al., 2017).

2.3.2. Voice Identification

The voice of an individual can be regarded as amongst the biometric feature due to two main reasons. Firstly, the voice is the outcome of the functioning of the physiological element which is called the voice tract. Moreover, it is the behavioral feature which is called the voice accent. By merging these factors, it isn't feasible to reproduce the voice of another person accurately. Voice recognition is the technology by which sound is transformed into electrical signals. These electrical signals are converted into coded patterns for verification. The phrases or words are recorded with a microphone. The specimen of input voice and registered models are associated to yield the ratio of likelihood (Khoh et al., 2019).

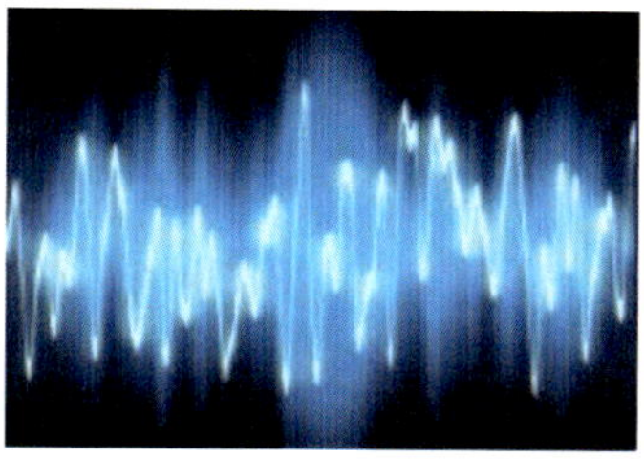

Figure 2.11. Voice signal recorded by a sensor.

Source: https://www.hindawi.com/journals/js/2018/9845321/.

Voice variation due to aging must also be well-thought-out by recognition systems. Moreover, unsanctioned users can record the voices of authorized clients and run them over the authentication process to attain user access control. To avert the probabilities of unsanctioned access with the aid of capturing devices, voice identification systems will ask customers to repeat several casual phrases which are given by the system throughout the authentication process.

2.4. FACTORS OF ASSESSMENT

As described in the former section, the range of biometric verification techniques is offered by several researchers. The extent of safety is amongst the main worries while assessing a specific technique. Several parameters exist by which the technique's performance can be measured. CER (Cross Error Rate), FRR (False Reject Rate), DET (Detection Error Tradeoff), FTC (Failure to Capture Rate) are some normally used factors (Malik et al., 2019).

2.5. BIOMETRIC IDENTIFICATION FRAMEWORK

The basis of the biometric identification system comprises two stages which can be exhibited in Figure 2.12 and defined as follows (Banerjee et al., 2018):

2.5.1. Stage of Enrolment

i. Attaining the subject

ii. Extraction of feature, i.e., salient has to be extracted.

iii. Storing these characteristics in the database.

2.5.2. Stage of Recognition

i. Attaining the subject from an individual.

ii. Extraction of a feature.

iii. Matching, which matches the characteristics set against the characteristics of every person on the database.

Though, in attaining the biometric feature, biometric signals experience many obstacles that might be generated by an individual or a sensor which generally leads to a reduction of quality and incorrect rate of recognition. As displayed in Figure 2.12, the signal isn't stable across the measurement. However, the main hindrance might be known as the source of intra-subject change like limitations of a sensor (η_s), inherent aging of biometric feature (η_a), change in the interaction of the user (η_v), an environment of acquisition (η_e), and some other factors impacts (η_o) (Bhattacharyya et al., 2009).

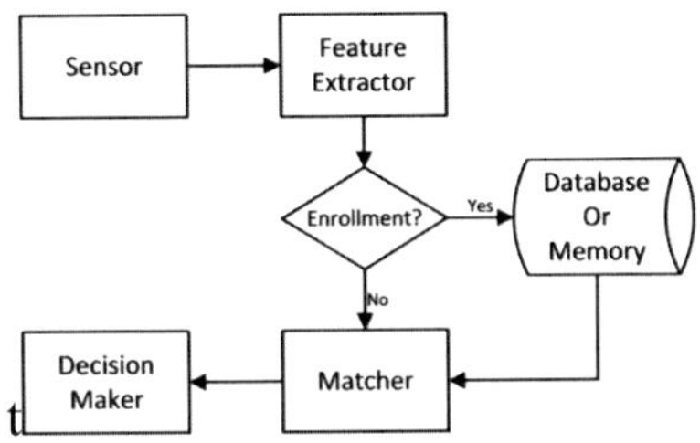

Figure 2.12. Basis of biometric identification system.

Source: *https://www.researchgate.net/figure/Modules-of-a-Biometric-System_fig1_322230928.*

2.6. CRITERIA OF CHOOSING BIOMETRIC FEATURE

For planning any biometric identification system, the confirmation of biometric features along with anticipated features need to be taken considered. Thus, an anticipated question is asked "How to select the biometric feature?" which might be well-thought-out as amongst the challenges on designing biometric identification systems (Syazana et al., 2016). The answer relies on the utilization of a system that might be anticipated, although, there is completely distinctiveness or individuality approved entirely in biometric features, searching for the highest rate of differentiating the biometric features for personal recognition. Though, as exhibited in the Figure underneath, there exists a set of properties or features that might fulfill the requirements of the designing system, like individuality or distinctiveness, performance, universality, and user acceptance.

2.7. MAIN RESEARCH CHALLENGES IN THE BIOMETRIC SYSTEMS

The main objective of biometric systems is to recognize the individuals precisely, which means that a biometric system must decrease the rate of error recognition. The best biometric system must have the lowest errors recognition rate. Though, such metrics are utilized by the researchers for measurement of error rate as FMR (False Match Rate), FNMR (False Non-Match Rate), FPIR (False Positive Identification Rate), and FNIR (False Negative Identification Rate) (Jaber & Younis, 2014).

For obtaining the optimum outcomes of the measurements above, the biometric system must have a suitable: (a) Sensor, (b) Characteristic representation scheme, and (c) Resemblance matcher. Two conditions should be fulfilled in order to minimize the recognition errors:

i. Intra-subject resemblance must increase. This is the resemblance between distinct samples for the similar biometric feature attained of the similar subject.

ii. Inter-subject resemblance must decrease. This is the resemblance between distinct samples of the biometric feature attained from diverse subjects.

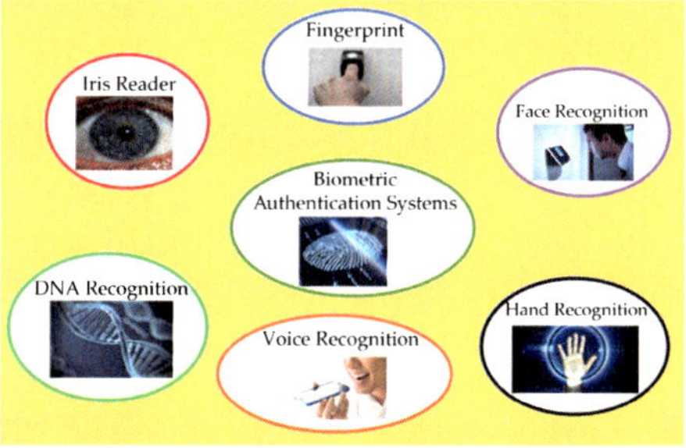

Figure 2.13. Biometric recognition system and its aims.

Source: *https://www.elprocus.com/biometric-authentication-system-applications/*

Additionally, biometric features can be collected under four categories of applications reliant on their usability and their rate of uniqueness and permanence for an individual (Das et al., 2018):

i. In the enforcement of law and forensic applications, DNA and palm print are mostly utilized.
ii. In security and commercial applications, the favored characteristics to utilize are voice, signature, vascular patterns, and hand geometry.
iii. In recognition of an individual, ear gait, electrocardiogram, keystroke dynamic, electroencephalogram signals are normally used.
iv. In the Management of security, fingerprint, face recognition, and iris are usually used.

Due to the diverse nature of biometric applications, there isn't any biometric feature that is an optimum option to fulfill the key requirements of all of the applications. The concept of fusion or amalgamation of two or more two biometric features might score the anticipated level of performance level. Normally such systems are known as multi-biometric systems (Joshi et al., 2015).

2.8. DEVELOPMENT OF BIOMETRIC IDENTIFICATION

This section provides the historical development of some significant features like fingerprint, iris, face, palm print, and hand geometry.

2.8.1. Historical Advancement in Fingerprint

"Possibly the most attractive and distinctive of all apparent marks are the tiny furrows with the prevailing ridges and the pores that are inclined in the singularly intricate but even order on under surfaces of the feet and hands."—Sir Francis Galton (Kaur & Madaan, 2015).

Fundamentally, the arrangement of fingerprint pictures can be: Plain/flat, Rolled/full, and Latent which are present but concealed, and might develop to become evident in the future. Plain and rolled fingerprint images can generally be obtained utilizing the live scan fingerprint sensor. The fingerprint traits can be classified into three levels as (a) Traits capture tiny details of fingerprint-like ridge flow, pattern type, singular point, and ridge frequency; (b) Traits refer to minutiae like ridge bifurcations and endings, and (c) Traits takes the dimensional characteristics of the ridge and comprises extended characteristics like deviation of ridge path, shape, width, incipient ridge, creases, and some other everlasting details.

The methods utilized in the fingerprint are: (i) image correlation, (ii) ridge traits matching, and (iii) minutia matching. Minutia-centered matching is the most commonly utilized technique in fingerprint due to two reasons: (i) minutia has been considered successful for fingerprint contrast by forensic inspectors for almost more than a hundred years, and (ii) storage effective to signify minutiae (Bhattacharyya et al., 2009).

B. **Historic Advancement of Face Recognition** "This acknowledgment problem is made difficult by the unusual instability in tilt and head pivot, angle and lighting intensity, outward appearance and maturing."—Woodrow Bledsoe.

The managing of matching of the face was done subsequently reliant on 20 standardized separations obtained from facial landmarks. In 1973 such kind of framework aimed to focus on such kind of facial landmarks, which generally have been exhibited the main mechanized face recognition framework.

C. **Historical Advancement in the Iris Identification** "For enthusiasms behind rapid and reliable distinct ID, it is often problematic to envision more qualified than the secured, unvarying, internal organ of an eye, that is quickly unmistakable distantly and that exposes uneven morphogenesis of the high statistical intricacy."—JohnDaugman.

The texture and color of an eye are used as an approach for identifying individuals and suggested by Bertillon. Truthfully, the iris is well-thought-out as amongst the best biometric features for such kinds of security applications. Conversely, in the iris identification system, a group of steps should be included as normalization of the image, trait extraction, and the classifier (Malik et al., 2014).

In the year 1936, it was suggested to use iris designs for human recognition. The 1st patent for the iris identification system was developed by (Ahmed et al., 2016). The patent comprises three main stages and these are capturing of image, extraction of feature, and matching. Whereas the main working iris identification framework was made and applied in the year and developed: (a) a camera to take the image of the iris, (b) an Image processing algorithm to process the image of an eye and develop the region of iris, and (c) The famous iris code representation to describe the images of iris as the binary code.

United Arab Emirates border control system was well-thought-out as the key substantial organizations of iris identification in 2001. In the year 2003, iris identification was used to streamline immigration control for habitual travelers at Amsterdam airport. Moreover, the iris recognition-centered immigration system operating at key airports in the United Kingdom for around 10 years, before this system was withdrawn from service in the year 2013. Between Canada and United States, an iris-centered border control system was utilized to quicken immigration endorsement for pre-confirmed travelers. Similarly, iris recognition was extensively used by the United States military to execute field tasks in Iraq and Afghanistan (Róka et al., 2007).

D. **Historical Advancement in Palmprint Identification:** Palmprint used for human recognition trails back to Chinese actions for sale in the 16th century. In 1684 offered dermatoglyphics, the examination of epidermal ridges, and the arrangement of these ridges on the surface of the palm. The key cautious capturing of a finger, hand, and palm images for differentiating proof was operated in the year 1858 discoursed the foundation of modern fingerprint science, and presented creases and palmar ridges. He recommended that the ridges on palms, fingers tips, and soles are unique and continuous. Galton described the individualities in ridges as the minutiae and familiarized numerous distinct minutiae kinds. The first utilization of palmprints in the criminal case happened in 1931 in the British court. The first programmed palmprint recognition system became accessible in the early 1990s (Fenker et al., 2013).

2.8.2. Historical Advancement in Ear Identification

Appearance, morphology, and structure of the human ear have generally been studied as the biometric feature, which provides various changes such as (a) Stable structure despite aging, particularly after fourth age, (b) Disturbed by variations in facial expression, and (c) capturing of image doesn't include apparent interaction with the sensor.

The performance of human ear identification algorithms has usually been verified in some of the standard ear sets of data, which outcomes in good identification precision under governed conditions. In any circumstance, the implementation of ear identification methods on the non-ideal image obtained under numerous illumination and obstruction situations is still to be set up. Several difficulties need to be defeated to make this possible (Bhattacharyya et al., 2009).

2.8.3. Historical Advancement in Gait Identification

For human recognition at distance, particularly for clandestinely identifying people in unrestricted conditions with a difficult subject. In this circumstance, the subject of interest, might not be liaising with the biometric system. In comparison to iris and fingerprint, gait should be observed in remain off parting of a biometric system which might only occur with substantial effort to be attained at a wide remain off parting (Bhattacharyya et al., 2009).

Consequently, gait-centered human identification has obtained some interest for biometric identification at some distance (Kaur & Madaan, 2015). Gait is categorized as the locomotion pattern in a living being. Human gait is how individuals walk. Whereas the proper meaning of gait relating to the movement of humans, algorithms for gait identification include both statistical and dynamic characteristics of the movement of the human body (Bhattacharyya et al., 2009).

2.8.4. Historical Advancement in Hand Geometry Identification

This refers to the geometric structure of the human hand, which can authenticate the individuals with the help of fingers width at several locations, the thickness of palm, width of the palm, and length of the fingers (Jain & Kumar, 2012).

The measurement of hand geometry is non-intrusive and the confirmation comprises the basic making of extracted characteristics.

Since 1970, hand geometry-centered verification systems are available, when the former studies on hand geometry biometrics are available as patent or application-based illustrations. In these particular applications, this system works as the verification approach. In addition to the shape of hand variations with time or aging. Conversely, more current researches discovered the utilization of hand geometry in combination with palm print and fingerprint in multibiometric configuration for enhanced accuracy (Bhattacharyya et al., 2009).

2.8.5. Historical Advancement in Periocular Identification

The Periocular is a region near the eyes. It comprises eye and skin eyebrows. As the biometric, this periocular region signifies the good trade-off amongst the complete face region or utilizing only iris for identification (Ross et al., 2006).

A specific image can usually be taken in the INR spectrum in order to decrease illumination discrepancy compared to the visible spectrum. The benefits of utilizing the periocular biometric feature are as:

i. In an iris, when the iris image isn't clear, or there exists low resolution, the nearby region might be utilized.
ii. This periocular region can provide data regarding eye shape that might be helpful as the delicate biometric.
iii. When some parts of the face associated with the nose and mouth are obstructed, the periocular region can be used to decide the character.

2.9. DISCUSSION AND CLARIFICATION OF UNSOLVED ISSUES

The biometric identification issues arise due to distinctiveness issues, permanence issues, the absence of matching and extraction methods, and also some application-specific issues. The biometric identification problems can normally be split into two categories, problems that are fundamentally associated with the design of the recognition system and the specific issues which are associated with applications that utilize biometric identification. In other words, there exist two major unsolved problems on biometric identification (Mbunge & Rugube, 2018):

i. Methods to protect the biometric system from assaults and provide assurances on the privacy of the user.
ii. Methods to examine the suitability of biometric systems and approximate the return to expenditure.

2.9.1. Distinctiveness

Determining the information level at which the uniqueness must be measured, is reflected as the basic problem in approximating the biometric feature individuality (Tatepamulwar & Pawar, 2014).

Distinctiveness can be defined centered on:

i. The biological feature (I(Y: B));

ii. The sensed images which are obtained from the subject (I(Y: M));

iii. The trait extracted from these sensed samples (I(Y:X)).

Along these particular lines, it is hard to compute the peculiarity of biological features, since one is dependent on these sensed samples which are available for investigation, and comprise of different types of noises assorted with the biometric feature information. Furthermore, the subject identification will be completely reliant on the characteristics extracted from these sensed images.

The absence of vigorous statistical models which depict the inter-and intra- subject discrepancies precisely is regarded as the primary problem in approximating the distinctiveness in biometric features centered on its characteristics representation. In this manner, approaching the entropy functions H(X), H(Y|X), or H(X|Y) turns into a hard job. The huge majority of the struggles made so far to assess the distinctiveness of biometric features requires simplifying expectations in order to keep the problem tractable (Pankanti et al., 2006).

The capacity of a biometric system to achieve low rates of error can be well-thought-out as evidence of the high individuality of underlying biometric features. This is proven that $H(Y|\hat{y})$ can be regarded as an upper bound on the function H (Y | X), where $H(Y|\hat{y})$ is the component of error rates of the biometric system. The approximation of distinction reliant on an observational system has some disadvantages, for instance:

i. As rates of error are dependent on a database, it isn't easy to induce them when the size of the population upsurges in large numbers.

ii. The resulting approximation is just the loose minimum on actual distinctiveness (Jain et al., 2006).

2.9.2. Perseverance of Biometric Features

Biometric feature persistence is connected to growing old which is mentioned to alteration in the biometric feature over the range of period, so it can affect the accuracy of a biometric system. However, there are two types of aging: template and trait aging. Trait aging refers to natural variation in the feature itself over the lifespan of an individual and template aging refers to variations in the biometric template of an individual after some time.

Whereas template aging is unconditionally recognized with trait aging, the characteristics extracted from the biometric feature can improve the impact of feature aging in template aging (Zhang et al., 2019).

Despite the point that the deduction of biometric trait varies from person to person, in this way, an analysis "is it possible to measure the variation level which particular template or trait is needed to encounter over the lifetime of a person?." The answer must systematically build the system sometimes and to confirm and revive the biometric template of the individual in order to collect and calculate the age-associated variations(Jain & Kumar, 2010).

2.9.3. Unrestricted Biometric Sensing Environment

The acquiring of biometric features is a hard task in the majority of the applications of individual identification. The sensing manner, acceptance of user, rotations, and obstruction are crucial problems that the process of acquisition face to get the subject image.

Table 2.1 gives a quick summary of the development of prevailing biometric traits identification systems.

Biometric trait	By	Year	Description of Development
Fingerprint	China	AD 600	Fingerprint to close legal documents and contracts
Palmprint	China	16th century	Utilized for human recognition follows back to Chinese activities for sale
Palmprint	Grew	1684	Presented the dermatoglyphics
Palmprint	Herschel	1858	The first efficient capture of a finger, palm, and hand image for recognition purposes
Fingerprint	Henry Faulds	1880	Article on fingerprint printed in Nature
Fingerprint	Juan Vocetich	1892	Utilization of paper and ink for fingerprint
Fingerprint	Argentina	1893	The first utilization of fingerprints as the proof for forensic
Fingerprint	Scotland Yard	1901	Implemented Galton system of classification
Face	Soviet Union	1915	35 millimeter still camera

Fingerprint	FBI, USA	1924	Fingerprint recognition division of Federal Bureau of Investigation
Palmprint	British court	1931	The first stated utilization of palmprints in the criminal case
Iris	Frank Burch	1936	The idea of utilizing iris identification patterns for human recognition
Fingerprint	Mitchell Trauring	1963	The first paper in the automatic fingerprint matching
Face	Woodrow Bledsoe	1964	Automated face recognition
Fingerprint	FBI	1970	Instigation of AFIS
Face	Takeo Kanade	1973	Thesis of FAR
Iris	Flom and Safir	1985	First iris identification patent
Iris	John Daugman	1989	First iris camera
Face	USA	1990	Surveillance camera
Fingerprint	USA		Capacitive and optical sensor
Palmprint	NSTC		First automated palmprint recognition system
Iris	John Daugman	1991	Iris identification patent
Face	Kodak		Digital camera
Face	Turk & Pentland		Eigen's face
Iris	Iris Scanner system	1995	Commercial iris camera
Face	Penev & Atick	1996	Local feature examination
Face	Wescott et al.	1997	Elastic Bunch Graph Matching
Fingerprint	Thomson-CSF		Swipe sensor
Face	Sharp	2000	Camera phone 320 pixels
Iris	UAE	2001	An established iris identification system for control of border crossing
Face	Viola & Jones		Face detector
Iris	USA	2002	Utilization of iris identification in field operation
Iris	Amsterdam, Netherlands	2003	Clearance of immigration in Schiphol airport

Iris	Security Metrics	2004	A portable iris identification device
Fingerprint	US Homeland Security		DHS US-VISIT
Fingerprint	TBS	2005	Touchless three-dimensional sensor
Iris	Sarnoff	2006	Iris identification device on the move
Face	Ahonen et al.		Local binary pattern
Fingerprint	US-VISIT	2007	Slap sensor for quick ten-print capture
Fingerprint	FBI	2008	NGI
Iris	India	2009	Utilizing iris in the main database for AADHAR ID
Face	Wright et al.		Sparse illustration
Fingerprint	INDIA		India began giving 12-digit UID number
Iris	Mexico	2010	National ID including the iris trait
Face	Microsoft Kinect		RGB-D camera Microsoft Kinect 480 pixels @30fps
Iris	John Daugman	2011	Patent finished
Iris	Indonesia		National ID comprising the iris feature
Face	Samsung		Galaxy nexus
Iris	AOptix	2013	Smartphones devise and application for iris identification
Iris	DeltaID		
Face	Google glass		Wearable cameras
Face	Jia et al.	2014	Deep network Caffe
Fingerprint	Apple Pay		2-factor verification
Fingerprint	SAFRAN		Touchless swipe sensor
Fingerprint	Apple		TouchID sensor
Face	Google & Intel	2015	Mobile phone REG-D camera
Face	NYPD, USA		Body camera utilized by NYPD

REFERENCES

1. Abraham, A., Jain, R., Thomas, J., & Han, S. Y. (2007). D-SCIDS: Distributed soft computing intrusion detection system. *Journal of Network and Computer Applications*, *30*(1), 81–98.
2. Adeoye, O. S. (2010). A survey of emerging biometric technologies. *International Journal of Computer Applications*, *9*(10), 1–5.
3. Ahmed, S. B., Razzak, M. I., & Alhaqbani, B. (2016, March). The minutiae based latent fingerprint recognition system. In *Proceedings of the International Conference on Internet of things and Cloud Computing*, 3, 1–9.
4. Aldeen, Y. A. A. S., Salleh, M., & Razzaque, M. A. (2015). A comprehensive review on privacy preserving data mining. *SpringerPlus*, *4*(1), 1–36.
5. Alrahawe, E. A., Humbe, V. T., & Shinde, G. N. (2019). An Analysis on Biometric Traits Recognition. *International Journal of Innovative Technology and Exploring Engineering (IJITEE) ISSN*, 3, 2278–3075.
6. Aslam, T. M., Tan, S. Z., & Dhillon, B. (2009). Iris recognition in the presence of ocular disease. *Journal of The Royal Society Interface*, *6*(34), 489–493.
7. Bailador, G., Sanchez-Avila, C., Guerra-Casanova, J., & de Santos Sierra, A. (2011). Analysis of pattern recognition techniques for in-air signature biometrics. *Pattern Recognition*, *44*(10–11), 2468–2478.
8. Banerjee, A., Basu, S., Basu, S., & Nasipuri, M. (2018). ARTeM: A new system for human authentication using finger vein images. *Multimedia Tools and Applications*, *77*(5), 5857–5884.
9. Bao, W., Huang, M., & Xiang, X. (2021). Enhancing Metric-Based Few-Shot Classification With Weighted Large Margin Nearest Center Loss. *IEEE Access*, *9*, 90805–90815.
10. Bhattacharyya, D., Ranjan, R., Alisherov, F., & Choi, M. (2009). Biometric authentication: A review. *International Journal of u-and e-Service, Science and Technology*, *2*(3), 13–28.
11. Burr, W. E., Dodson, D. F., & Polk, W. T. (2006). *Information security: Electronic authentication guideline nist*. Technical report, *Tech. Rep. Special Rep,* *4*(3), 800–63.
12. Chihaoui, M., Elkefi, A., Bellil, W., & Ben Amar, C. (2016). A survey of 2D face recognition techniques. *Computers*, *5*(4), 21.

13. Chowhan, S. S., & Shinde, G. N. (2011). Iris recognition using fuzzy min-max neural network. *International Journal of Computer and Electrical Engineering*, *3*(5), 743.
14. Das, R., Piciucco, E., Maiorana, E., & Campisi, P. (2018). Convolutional neural network for finger–vein-based biometric identification. *IEEE Transactions on Information Forensics and Security*, *14*(2), 360–373.
15. de-Santos-Sierra, A., Sánchez-Avila, C., Del Pozo, G. B., & Guerra-Casanova, J. (2011). Unconstrained and contactless hand geometry biometrics. *Sensors*, *11*(11), 10143–10164.
16. Depren, O., Topallar, M., Anarim, E., & Ciliz, M. K. (2005). An intelligent intrusion detection system (IDS) for anomaly and misuse detection in computer networks. *Expert systems with Applications*, *29*(4), 713–722.
17. Duta, N. (2009). A survey of biometric technology based on hand shape. *Pattern recognition*, *42*(11), 2797–2806.
18. Fancourt, C., Bogoni, L., Hanna, K., Guo, Y., Wildes, R., Takahashi, N., & Jain, U. (2005, July). Iris recognition at a distance. In *International Conference on Audio-and Video-Based Biometric Person Authentication*, *3*(2), 1–13.
19. Fang, Y., Kang, W., Wu, Q., & Tang, L. (2017). A novel video-based system for in-air signature verification. *Computers & Electrical Engineering*, *57*, 1–14.
20. Farnaaz, N., & Jabbar, M. A. (2016). Random forest modeling for network intrusion detection system. *Procedia Computer Science*, *89*, 213–217.
21. Fenker, S. P., Ortiz, E., & Bowyer, K. W. (2013). Template aging phenomenon in iris recognition. *IEEE Access*, *1*, 266–274.
22. Gafurov, D., Helkala, K., & Søndrol, T. (2006). Biometric Gait Authentication Using Accelerometer Sensor. *J. Comput.*, *1*(7), 51–59.
23. Guerra-Casanova, J., Sánchez-Ávila, C., de Santos Sierra, A., & Del Pozo, G. B. (2011). Score optimization and template updating in a biometric technique for authentication in mobiles based on gestures. *Journal of Systems and Software*, *84*(11), 2013–2021.
24. Guest, R., Brockly, M., Elliott, S., & Scott, J. (2016). An assessment of the usability of biometric signature systems using the human-biometric sensor interaction model. *International Journal of Computer Applications in Technology*, *53*(4), 336–347.

25. Gul, I., & Hussain, M. (2011). Distributed cloud intrusion detection model. *International Journal of Advanced Science and Technology, 34*(38), 135.
26. Hashim, N. A., Abidin, Z. Z., & Shibghatullah, A. S. (2017). Iris Feature Detection using Split Block and PSO for Iris Identification System. *Journal of Telecommunication, Electronic and Computer Engineering (JTEC)*, *9*(1–2), 99–102.
27. Hayday, A. C. (2000). γδ cells: a right time and a right place for a conserved third way of protection. *Annual review of immunology, 18*(1), 975–1026.
28. Huang, M., Chen, Y., Chen, B. W., Liu, J., Rho, S., & Ji, W. (2016). A semi-supervised privacy-preserving clustering algorithm for healthcare. *Peer-to-Peer Networking and Applications*, *9*(5), 864–875.
29. Huang, M., Xiang, X., Chen, Y., & Fan, D. (2018). Weighted large margin nearest center distance-based human depth recovery with limited bandwidth consumption. *IEEE Transactions on Image Processing*, *27*(12), 5728–5743.
30. Jaber, Z. Q., & Younis, M. I. (2014). Design and implementation of real time face recognition system (RTFRS). *International Journal of Computer Applications*, *94*(12), 1–18.
31. Jabez, J., & Muthukumar, B. (2015). Intrusion Detection System (IDS): Anomaly detection using outlier detection approach. *Procedia Computer Science*, *48*, 338–346.
32. Jain, A. K., & Feng, J. (2008). Latent palmprint matching. *IEEE Transactions on Pattern Analysis and Machine Intelligence*, *31*(6), 1032–1047.
33. Jain, A. K., & Kumar, A. (2010). Biometrics of next generation: An overview. *Second Generation Biometrics*, *12*(1), 2–3.
34. Jain, A. K., & Kumar, A. (2012). Biometric recognition: an overview. *Second generation biometrics: The Ethical, Legal and Social Context, 3*, 49–79.
35. Jain, A. K., Nandakumar, K., & Ross, A. (2016). 50 years of biometric research: Accomplishments, challenges, and opportunities. *Pattern Recognition Letters*, *79*, 80–105.
36. Jain, A. K., Ross, A., & Pankanti, S. (2006). Biometrics: a tool for information security. *IEEE Transactions on Information Forensics and Security*, *1*(2), 125–143.

37. Joshi, M. P., Uppal, R. S., & Kaur, L. Development of Vision Based Iris Recognition System. (2015). *International Journal of Advanced Engineering Sciences and Technologies*, 6, 277–281.
38. Kaur, R., & Madaan, S. (2015). IRIS recognition by using HCT, PCA, hamming distance. *International Journal*, *2*(3), 11–16.
39. Kaur, S. (2011). Speaker Verification using LabVIEW. *International Journal of Computer Applications*, *21*(4), 8.
40. Khoh, W. H., Pang, Y. H., & Teoh, A. B. J. (2019). In-air hand gesture signature recognition system based on 3-dimensional imagery. *Multimedia Tools and Applications*, *78*(6), 6913–6937.
41. Kumar, A., Wong, D. C., Shen, H. C., & Jain, A. K. (2003, June). Personal verification using palmprint and hand geometry biometric. In *International Conference on Audio- and Video-Based Biometric Person Authentication*, 4, 668–678.
42. Lee, D. W., Gardner, R., Porter, D. L., Louis, C. U., Ahmed, N., Jensen, M., ... & Mackall, C. L. (2014). Current concepts in the diagnosis and management of cytokine release syndrome. *Blood, The Journal of the American Society of Hematology*, *124*(2), 188–195.
43. Lee, S. X., Leemaqz, K. L., & McLachlan, G. J. (2019). PPEM: Privacy□preserving EM learning for mixture models. *Concurrency and Computation: Practice and Experience*, *31*(24), 5208.
44. Li, F., Ma, J., & Li, J. H. (2009). Distributed anonymous data perturbation method for privacy-preserving data mining. *Journal of Zhejiang University-Science A*, *10*(7), 952–963.
45. Li, Z., Yang, L., & Li, Z. (2019). Mixture-Model-Based Graph for Privacy-Preserving Semi-Supervised Learning. *IEEE Access*, *8*, 789–801.
46. Lu, Y., Wu, S., Fang, Z., Xiong, N., Yoon, S., & Park, D. S. (2017). Exploring finger vein based personal authentication for secure IoT. *Future Generation Computer Systems*, *77*, 149–160.
47. Malik, J., Elhayek, A., & Stricker, D. (2019). WHSP-Net: A weakly-supervised approach for 3D hand shape and pose recovery from a single depth image. *Sensors*, *19*(17), 3784.
48. Malik, J., Elhayek, A., Ahmed, S., Shafait, F., Malik, M. I., & Stricker, D. (2018). 3dairsig: A framework for enabling in-air signatures using a multi-modal depth sensor. *Sensors*, *18*(11), 3872.

49. Malik, J., Girdhar, D., Dahiya, R., & Sainarayanan, G. (2014). Reference threshold calculation for biometric authentication. *International Journal of Image, Graphics and Signal Processing*, *6*(2), 46.
50. Masmoudi, A., Puech, W., & Bouhlel, M. S. (2010). A new joint lossless compression and encryption scheme combining a binary arithmetic coding with a pseudo random bit generator. *International Journal of Computer Science and Information Security*, *8*(1), 170–175.
51. Mbunge, E., & Rugube, T. (2018). A robust and scalable four factor authentication architecture to enhance security for mobile online transaction. *International Journal of Scientific & Technology Research*, *7*(3), 139–143.
52. Nagar, A., Nandakumar, K., & Jain, A. K. (2010). A hybrid biometric cryptosystem for securing fingerprint minutiae templates. *Pattern Recognition Letters*, *31*(8), 733–741.
53. Nandakumar, K., & Jain, A. K. (2015). Biometric template protection: Bridging the performance gap between theory and practice. *IEEE Signal Processing Magazine*, *32*(5), 88–100.
54. Onesimu, J. A., Karthikeyan, J., & Sei, Y. (2021). An efficient clustering-based anonymization scheme for privacy-preserving data collection in IoT based healthcare services. *Peer-to-Peer Networking and Applications*, *14*(3), 1629–1649.
55. Osterburg, J. W., Parthasarathy, T., Raghavan, T. E. S., & Sclove, S. L. (1977). Development of a mathematical formula for the calculation of fingerprint probabilities based on individual characteristics. *Journal of the American Statistical Association*, *72*(360a), 772–778.
56. Ouda, M. A., Salem, S. A., Ali, I. A., & Saad, E. S. M. (2012). Privacy-preserving data mining (ppdm) method for horizontally partitioned data. *International Journal of Computer Science Issues (IJCSI)*, *9*(5), 339.
57. Paiva, J. S., Dias, D., & Cunha, J. P. (2017). Beat-ID: Towards a computationally low-cost single heartbeat biometric identity check system based on electrocardiogram wave morphology. *PloS One*, *12*(7), 0180942.
58. Pankanti, S., Prabhakar, S., & Jain, A. K. (2002). On the individuality of fingerprints. *IEEE Transactions on Pattern Analysis and Machine Intelligence*, *24*(8), 1010–1025.

59. Park, U., Jillela, R. R., Ross, A., & Jain, A. K. (2010). Periocular biometrics in the visible spectrum. *IEEE Transactions on Information Forensics and Security*, *6*(1), 96–106.
60. Rath, S. K., & Rautaray, S. S. (2014). A survey on face detection and recognition techniques in different application domain. *International Journal of Modern Education and Computer Science*, *6*(8), 34.
61. Róka, A., Csapó, Á., Reskó, B., & Baranyi, P. (2007). Edge detection model based on involuntary eye movements of the eye-retina system. *Acta Polytechnica Hungarica*, *4*(1), 31–46.
62. Ross, A. A., Jain, A. K., & Nandakumar, K. (2006). Levels of fusion in biometrics. *Handbook of Multibiometrics*, 2(1), 59–90.
63. Ross, A., Jain, A., & Pankati, S. (1999, March). A prototype hand geometry-based verification system. In *Proceedings of 2nd Conference on Audio and Video Based Biometric Person Authentication*, *5*, 166–171.
64. Saini, K., & Dewal, M. L. (2010). Designing of a Virtual System with Fingerprint Security by considering many Security threats. *International Journal of Computer Applications*, *3*(2), 25–31.
65. Shaheed, K., Liu, H., Yang, G., Qureshi, I., Gou, J., & Yin, Y. (2018). A systematic review of finger vein recognition techniques. *Information*, *9*(9), 213.
66. Shao, Y., Hong, W., & Li, Z. (2019). A new method to compute ratio of secure summations and its application in privacy preserving distributed data mining. *IEEE Access*, *7*, 20756–20766.
67. Sharma, S., & Ahuja, S. (2019). Privacy preserving data mining: A review of the state of the art. *Harmony Search and Nature Inspired Optimization Algorithms*, 3, 1–15.
68. Shelke, M. P. K., Sontakke, M. S., & Gawande, A. D. (2012). Intrusion detection system for cloud computing. *International Journal of Scientific & Technology Research*, *1*(4), 67–71.
69. Siddiqui, M. F., Siddique, W. A., Ahmedh, M., & Jumani, A. K. (2020). Face detection and recognition system for enhancing security measures using artificial intelligence system. *Indian Journal of Science and Technology*, *13*(09), 1057–1064.
70. Sondhiya, R., Shreevastav, M., & Mishra, M. (2013). To improve security in cloud computing with intrusion detection system using neural network. *International Journal of Soft Computing and Engineering (IJSCE) ISSN*, *4*, 2231–2307.

71. Suganthy, M., Ramamoorthy, P., & Krishnamoorthy, R. (2012). Effective Iris Recognition For Security Enhancement. *International Journal of Engineering Research and Applications (IJERA)*, *2*(2), 1016–1019.
72. Syazana-Itqan, K., Syafeeza, A. R., Saad, N. M., Hamid, N. A., & Saad, W. H. B. M. (2016). A review of finger–vein biometrics identification approaches. *Indian J. Sci. Technol*, *9*(32), 1–9.
73. Tatepamulwar, C. B., & Pawar, V. P. (2014). Comparison of biometric trends based on different criteria. *Asian Journal of Management Sciences*, *2*(3), 159–165.
74. Truta, T. M., Campan, A., & Sun, X. (2012). An overview of P-sensitive k-anonymity models for microdata anonymization. *International Journal of Uncertainty, Fuzziness and Knowledge-Based Systems*, *20*(06), 819–837.
75. Uludag, U., Ross, A., & Jain, A. (2004). Biometric template selection and update: a case study in fingerprints. *Pattern Recognition*, *37*(7), 1533–1542.
76. Walker, D. R. (2012). Biometric Technology in Law Enforcement. *Neurosurgery*, *71*(2), 197–200.
77. Xi, X., Yang, L., & Yin, Y. (2017). Learning discriminative binary codes for finger vein recognition. *Pattern Recognition*, *66*, 26–33.
78. Zarrabi, A., & Zarrabi, A. (2012). Internet intrusion detection system service in a cloud. *International Journal of Computer Science Issues (IJCSI)*, *9*(5), 308.
79. Zhang, W. K., & Kang, M. J. (2019). Factors affecting the use of facial-recognition payment: An example of Chinese consumers. *Ieee Access*, *7*, 154360–154374.

Chapter 3

Soft Biometrics for Human Recognition

CONTENTS

3.1. INTRODUCTION

Citing security concerns, the attention has shifted to multi-biometrics. Soft biometrics are ancillary data obtained from primary biometric (body and face) characteristics like gender, facial measurements, ethnicity, height, and color of skin (Dantcheva et al., 2010). They can be used to enhance the overall system efficiency and speed of a primary biometric system (that is, fuse face with marks) or to create human semantic analysis descriptions of a person. While using ethnicity and gender (e.g., elderly African man having blue eyes) in a fusion structure, this also limits the search to the entire dataset. The chapter presents a comprehensive overview of soft biometrics, concentrating on facial soft biometrics, and addresses a few of the characteristics of suggested extraction and classification algorithms, as well as their merits and drawbacks.

Security concerns become much more crucial and essential as our modem lives are becoming more automated. "is this the authorized individual to conduct such activity?," and "will this individual belong to this country?," "Is this the correct individual to be able to use the system?" all are queries we address in our everyday lives (Jaramillo & Zhang, 2013). There are two techniques for responding to these queries: one depending on "whatever you have" and referred to as knowledge factors, like ID cards, and the other depending on "what you know" and referred to as (ownership factors), like passwords (see Figure 3.1). These techniques, though, can be duplicated, stolen, or taken, necessitating the use of multiple IDs and a large number of passwords. As previously documented, breaches of card-based and password-based security policies cost governments, banks, and telecommunication firms millions of dollars each year (Noor et al., 2009). Biometrics is an open field for resolving the person identification problem.

Figure 3.1. Sets of information.

Source: *https://www.intechopen.com/chapters/60675.*

Biometrics depend on what is known as inherence factors to distinguish between legal and unlawful individuals (Clarke, 1994). The below are some of the benefits of biometric features (Koppen et al., 2006):

- They are one-of-a-kind for each person.
- They can't be readily observed, forgotten, borrowed, stolen, or shared; They always vary and are always available.
- They are always available and changing.
- They are difficult to convey to another person.

It's nearly difficult to deceive a biometric-based surveillance system. The term biometric is a combination of the Greek words bios, which means life, and metron, which means measurement. Biometrics is a research area that concentrates on analyzing and measuring an individual's specific traits to recognize or confirm a person's identification, and it is an important daily duty for a privacy system to ensure that services are only accessible by authorized users (Abdelwhab & Viriri, 2018).

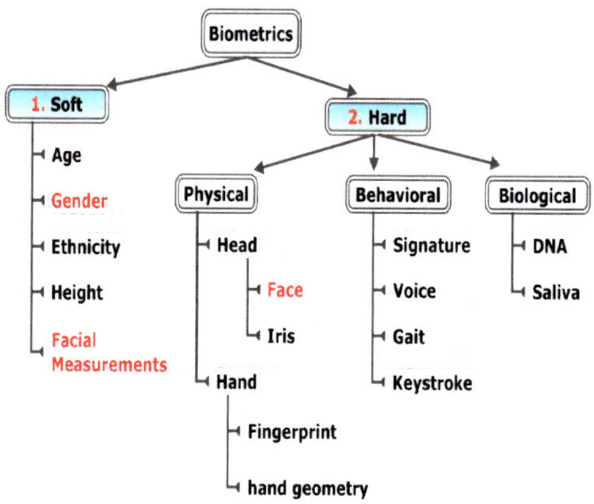

Figure 3.2. Types of biometrics.

Source: https://www.bookdepository.com/Machine-Learning-Biometrics-Jucheng-Yang/9781789235906.

It is separated into three categories: soft, traditional, and primary biometrics, as illustrated: As seen in Figure 3.2, traditional biometrics cope with biological, physical, and behavioral traits such as fingerprints, facial features, signature, eye, voice, gait, and DNA. Soft biometrics are associated with ancillary traits such as height, skin color, ethnicity, scars, and gender that provide little data but are insufficient to recognize a person

(Jain & Verma, 2012). To be identified as a biometric trait, physiological and behavioral human aspects must meet the following criteria (Nandakumar & Jain, 2015):

1) Acceptable: readily available as necessary.
2) The trait is universal: it exists in everyone.
3) Distinctive: can be used to distinguish between individuals.
4) Resistance to circumvention: cheating is difficult.
5) Collectible: the attribute is simple to gather and quantify.
6) They are permanent: they do not alter over time.

Moreover, no single biometric characteristic currently exists that satisfies all of the aforementioned characteristics, so no current biometric system can offer additional accurate foolproof identification. As a consequence, there is indeed a disparity for enhancing the identification quality and precision of primary biometrics utilizing soft biometrics (Chiu et al., 2018). This chapter is organized into the following sections: Section 3.2 discusses the advantages of soft biometrics, the limitations of unimodal biometric systems and also how multimodal biometric systems resolve these limitations, the importance of biometric fusion for overall effectiveness and measurement. Section 3.3 presents a comprehensive assessment of relevant publications, whereas Section 3.4 focuses on biometrics and Section 3.5 concludes the study.

3.2. SOFT BIOMETRICS

Soft biometrics offer further data that are not unique and persistent, therefore they cannot be used to reliably identify people. Yet, such ancillary data can be applied as a supplement to primary biometric characteristics (iris, facial, etc.), and all these aspects can be classified as physicality (e.g., ethnicity, skin color, gender), accessories (e.g., cap, eyeglasses) or clothing (e.g., garment color) (Park et al., 2010).

3.2.1. Advantages of soft biometrics

- it can be utilized to enhance a primary biometric system's identification speed and accuracy (Dantcheva et al., 2015).
- Can be employed when collecting a major biometric feature is difficult, the gathered data is unclear owing to sensor inaccuracy, or information is recorded from a distance without any user participation.

- Acceptable: data collection for identification does not necessitate collaboration among the sensor and the person, and it is easily accessible.
- Soft facial biometrics are cheap to calculate because they may be collected at a very similar time as primary face biometrics.
- Registering a person requires no collaboration and can be done at a distance; even system training can be done online.
- Because they contain a semantic meaning and may be comprehended by the human being as just a short and old African male, soft biometrics overcome the barrier between humans and machines.
- Since soft biometrics offer ancillary features and are not entirely distinct from short and old males, so they do not raise privacy concerns about storing and maintaining data.
- Indexing and filtering a vast database to reduce the number of data searches based on associated human attributes, such as feminine gender (Dantcheva et al., 2105).

3.2.2. System of Biometric

This is a vital pattern identification system that utilizes human attributes to recognize the person separated into the unimodal system while utilizing only one character and multi-biometric system so if utilizing multiple characteristics (Zewail et al., 2004). When trying to establish a credible biometric system, there are a few issues which require to be balanced and analyzed as required (Jain et al., 2004):

- No harm to users, according to a study firm that implanted a SIM card there beneath the skin for verification.
- Performance, defined as the maximum identification rate and system performance while tolerating the system's time invariance, external influences, and stability.
- Acceptability: Are individuals willing to put your biometric attribute to use?
- Circumvention refers to how readily your system can be bypassed or prevented through the use of deceptive approaches.
- Simple to use and accessible.

Since the collaboration with the consumer is required to gather the information, Unimodal systems suffering from a shortage of data due to the sensor or an individual, which can result in a poor identification rate, large failures to register rate, and a lack of people coverage area. So getting very high identification rates with a unimodal system is almost impossible; to boost the identification rate, we have to collect greater than one feature from the same sensor or several sensors, however as the identification rate rises, so do the complication and processing time (Zewail et al., 2004).

A few of the issues connected with unimodal biometric systems can be addressed by using multi-biometric systems that incorporate data from various sources (Khalifa & BenAmara, 2012). Yet, due to the requirement for further processing needs, higher-quality sensors, and enormous storage space, such a system has two key restrictions: firstly, the total cost of construction can be prohibitive. Secondly, the system demands more time for validation, which causes consumers to be inconvenienced. Soft biometrics, on the other hand, is a cost-cutting option that uses an identical sensor (Nandakumar & Jain, 2015). As indicated in Figure 3.3, the key stages for a biometric system include the following (Frischholz & Dieckmann, 2000):

- Registration is the first phase, in which the sensor collects the person's biometric features and saves them to the database as a blueprint for verification and eventual identification. The next stages require valid biometric registration.
- Using preprocessing techniques such as equalization of the histogram and cutting the region of concern to cope with luminance, enhance the saved data to achieve a high identification rate.
- Ability to extract attributes vector from the people for identification and matching this with the recorded template data.
- Template dataset: registering data entails putting biometric information in a set of data as a template that can be compared to.
- Classification and matching: biometric attribute data is compared to the dataset's template data. The comparable resemblance score or the cutoff value might be used to reject or approve a decision (Velardo & Dugelay, 2011).

The biometric system has two modes of operation: verification and identification modes. The verification method compares the individual to his template collected in the set of data, whereas the identification method compares the individual to all of the templates collected in the set of data (Tiwari et al., 2012).

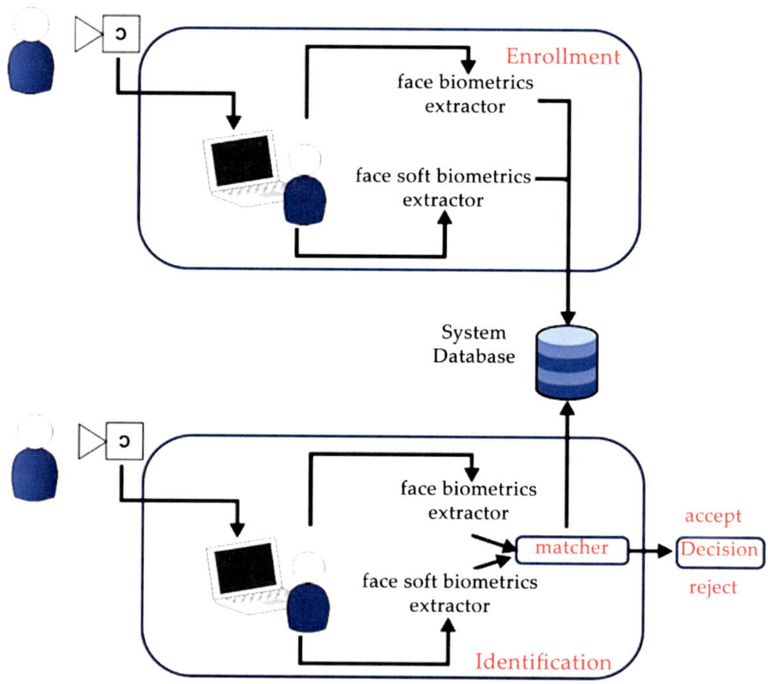

Figure 3.3. Identification and Enrollment schematic for biometric systems.

Source: *https://www.researchgate.net/publication/327293497_A_Survey_on_Soft_Biometrics_for_Human_Identification.*

3.2.3. Biometric fusion

We surmount the unimodal constraint by trying to minimize one or more of the acceptance and rejection error rates, as biometric data be affected by climatic conditions and may change with time, so by combining more than one characteristic or the same characteristic from more than one origin depending on the system criteria, as illustrated in Figure 3.4 (Algarni et al., 2020). Furthermore, there is no single best biometrics because different applications necessitate different policies, like border control, national identity cards, and distance learning, all of which necessitate enrollment failure rate and a low false accept rate. Fusion, on the other hand, is critical for increasing recognizing rates and can be done at various stages – classification stage, sensor, feature extraction, decision.

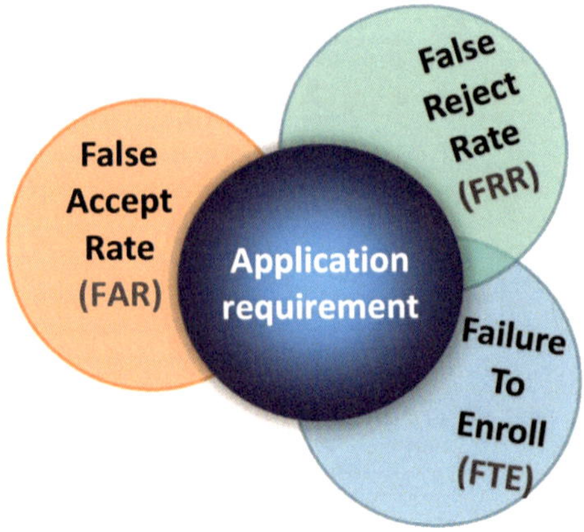

Figure 3.4. Performance analysis.

Source: https://www.researchgate.net/publication/327293497_A_Survey_on_Soft_Biometrics_for_Human_Identification.

As exhibited in Figure 3.5, Paliwal et al. (2002) classify fusion into two groups: pre-matching and post-matching.

- Pre-classification fusion (Ross & Poh, 2009): integration can be performed in two ways before the classification level:

1. Feature-level fusion: in feature fusion, we have a variety of data by combining different characteristics derived from captured pictures into a single set of features. As a result, feature sets must be transformed, fine-tuned, compressed, and normalized. In practice, feature-level fusion is difficult to obtain since concatenating different characteristics can cause dimensionality issues.
2. Sensor level: integrating raw information is difficult as it contains many non-important characteristics in addition to the area of interest, and data obtained from detectors can be noisy due to non-uniform luminous. Raw data collected employing various sensors or various snapshots of a biometric achieved utilizing a single sensor are referred to as sensor fusion. Face photos gathered from a variety of sources with varying resolutions may not be able to be combined (Batool & Chellappa, 2015).

- Post-classification fusion (Ho et al., 1994): there are three types of post-classification integration:

1. At the score level, scores are combined to create a single score value, which is then used to make decisions based on the baseline value. Because there is a spectrum that can be configured to increase and reduce false rejected and acceptance rates, the threshold makes the process more accurate than correct and incorrect. A lower limit, on the other hand, reduces the rate of inaccurately rejected applications while increasing the rate of inaccurately accepted applications (Lam & Suen, 1997).

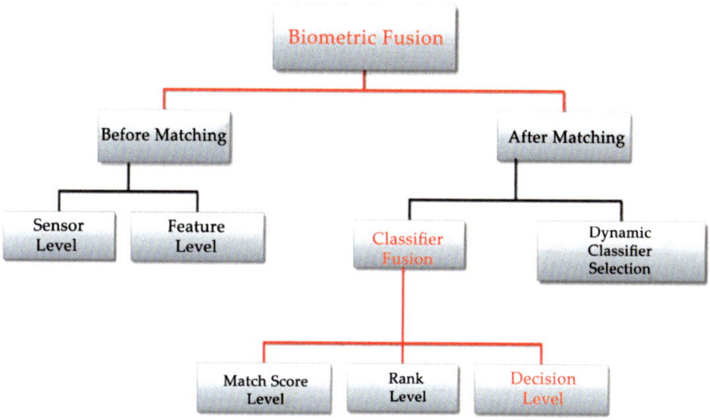

Figure 3.5. Levels of fusion.

Source: https://www.intechopen.com/chapters/60675.

2. At the rank level: the score values are organized in going down the order, indicating the potential of placing the most desired classes at the highest point of the list and the least desired classes at the bottom (Achermann & Bunke, 1996).
3. The decision level is entirely based on the score level's result value, and a final choice is made as to whether the identified person is a fraud to deny or an extraordinary to admit. Each classifier presents a difficult choice. The following options can be used to integrate the choices:

- Logic operator (and, or):

  The and-operator ensures that all classifiers produce identical results, whether rejecting or accepting, which is beneficial when lower false acceptance is needed. When a lower false rejection is needed, the or operator comes in handy.

- Fuzzy logic (Zadeh, 1996):

  Rather than rejection and acceptance, we have a truth-value that is somewhere in the middle.

- Majority voting:

  When a large proportion of the classifiers agree on a judgment, it is called a decision. To make sure that a judgment is made, we need more classifiers than there are classes.

3.2.4. Performance Analysis

A biometric system must be assessed and verified; while some assessment concepts include false acceptance rate, false rejection, and equal (Siff, 2017):

- FAR refers to the number of people who do not have authorization but are falsely accepted by the system as authorized individuals.
- FTE – Failure to enroll: refers to the percentage of people who are unable to register in the structure.
- EER stands for equitable rate false reject and false accept, and the higher the EER, the more precise the process. The FRR stands for the percentage of users who are allowed but are incorrectly rebuffed by the process.
- FTC – Failure to capture: the biometric characteristics were displayed correctly, but the framework was unable to detect them correctly.
- FRR: the authorized person's number was incorrectly rejected by the structure.

FRR = (number of false rejected/NAA) × 100%	(1)
F AR = (number of false accepted/NIA) × 100%	(2)

The terms NIA and NAA refer to the number of impostor attempts and authorized attempts, respectively. Some variables can affect the performance measurements, recognition rate, and accuracy of a biometric system (Siff, 2017):

- User wishes and willingness: because users are not required to interact with the system regularly, the system becomes infected, and accuracy suffers.
- Patients who have had plastic surgery or who do not have a hand cannot use a fingerprint.

- Low accuracy is caused by environmental variables such as rain humidity, steam, and high temperature. As you get older and perform better, your features alter.

Because the above variables influence all evaluation rates, any biometric system must measure error variables and tune and normalize them according to structure nature and prerequisites (Shen & Tan, 1999).

3.3. SOFT BIOMETRICS CURRENT TRENDS

In 1896, Alphonse Bertillon was the first to propose an anthropometric, morphological, and biometric personal identification system depending on hair, eye, and skin color. Facial recognition is less distinctive and more appropriate than iris recognition, but it is even now client-friendly, and individuals are more inclined to use it than other methods. The soft biometric system is categorized into three parts (Tome et al., 2014):

- Head attributes are where the investigation is currently focused due to the wealth of information available in this body part, such as hair and skin color and facial measurements.
- A person's weight and height are described as fat or tall using body characteristics.
- For a set of data indexing, global characteristics such as sex and ethnicity are used, which are fixed for the rest of one's life.

  As illustrated in Table 3.1, soft biometric characteristics can also be classified based on their distinctiveness and permanence. Because ethnicity and gender do not change with time, the constancy of an attribute demonstrates its strength over time. The capacity of a characteristic to distinguish between individuals is referred to as distinctiveness (Achermann & Bunke, 1996).

Table 3.1. Soft biometric Traits of the Face

Soft biometric traits	Face	Permanence	Distinctiveness
Facial measurements	Face	High	Medium
Gender	Face	High	Low
Skin color	Face	Medium	Low
Eye color	Face	Medium	Medium
Tatoo	Face	High	High

Age	Face	Low	Medium
Mustache	Face	Low	Low

The soft biometric characteristics of the head are the subject of this paper. Faces can conveniently identify human beings since they do not alter over time or in a wide range of ways. Facial features provide different information when clipped, resized, and shown from different sides, according to Lin (Almudhahka et al., 2017).

The related works section displays some of the significant works from 2000 to 2017 in chronological order.

The fathers of soft biometrics, Jain et al. (2004), presented it as ancillary data, but they are unable to individually verify the person due to a lack of permanence and distinctiveness. To strengthen the primary fingerprint system, they suggest using demographic data like height, gender, and ethnicity as soft biometrics. Findings demonstrate that utilizing soft biometrics improved fingerprint identification performance by 5%.

The experimental finding of Almudhahka et al. (2017) demonstrates that soft biometrics can be utilized as secondary data to enhance primary biometrics and can be obtained from a distance; fusion is performed at the scoring stage. Three feature extraction techniques are used by Park and Jain et al. (2010):

- Extraction of facial characteristics such as the eyes and nose using an active appearance method;
- Morphological operators;
- Morphological operators.

The system is evaluated using two sets of data. They found that using soft biometrics such as facial marks, gender, and ethnicity to improve recognition rates. When facial images are partially damaged, soft biometric characteristics can be regarded as a substitute. Because a person's ethnicity and gender do not change with time, they can be utilized to cleanse the dataset and narrow the search list. Even so, as quality improved, so did complexity, and facial mark removal is dependent on picture resolution and the regulated atmosphere required (Tome et al., 2014).

3.4. FUTURE WORK AND CHALLENGES

Multi-biometric methods collect various characteristics from various sensors to surmount the constraints of uni-modal biometric. Even so, lengthening the processing time and raising the number of verification steps, like a system will reduce performance, causing problems for users. As a result, we fuse primary and soft biometrics to enhance the overall efficiency of the primary biometric system to create a credible and customer-friendly biometric system (Zhang et al., 2019).

Soft biometrics are non-intrusive and computationally efficient, allowing for quick, pose-invariant, and enrolment-free biometric analysis. Even so, since soft biometric traits alter over time and lack uniqueness, biometric systems depending on them alone cannot give precise recognition, so there are still many obstacles in this zone. Fuzzy logic can help with variable adjustments, such as decision thresholds and fusion rules because otherwise, the error rate will rise.

Because soft biometrics are delicate to lighting, pose variations, and expression variations, we can use profound learning for attribute pre-processing and extraction. New soft biometric characteristics, such as face distance measurement and the ratio of head to body size, can also be presented (Denman et al., 2009).

We discovered that there is no single best biometric technique for user identification after conducting a comprehensive survey on soft biometrics. For instance, in protection, a zero false acceptance rate is required, and the false rejection rate must be reduced; however, in civilian applications, the opposite is required, so any biometric system must strike an equilibrium between complexity and authentication reliability. As a consequence, conventional biometrics have a low recognition rate since they require the client's coordination, operate in a regulated environment, and pose a privacy risk. The approach is to use multi-biometrics, but the structure still suffers from high computation costs and lengthy processing steps. A further option is to utilize soft biometrics to expand the demographic coverage while lowering the platform's complexity and cost (Denman et al., 2015).

REFERENCES

1. Abdelwhab, A., & Viriri, S. (2018). A survey on soft biometrics for human identification. *Machine Learning and Biometrics, 37*, 1–8.
2. Achermann, B., & Bunke, H. (1996, August). Combination of face classifiers for person identification. In *Proceedings of 13th International Conference on Pattern Recognition, 3,* 416–420.
3. Algarni, A. D., El Banby, G. M., Soliman, N. F., El-Samie, F. E. A., & Iliyasu, A. M. (2020). Efficient implementation of homomorphic and fuzzy transforms in Random-Projection encryption frameworks for cancellable face recognition. *Electronics, 9*(6), 1046
4. Almudhahka, N. Y., Nixon, M. S., & Hare, J. S. (2017). Semantic face signatures: Recognizing and retrieving faces by verbal descriptions. *IEEE Transactions on Information Forensics and Security, 13*(3), 706–716.
5. Batool, N., & Chellappa, R. (2015). Fast detection of facial wrinkles based on Gabor features using image morphology and geometric constraints. *Pattern Recognition, 48*(3), 642–658.
6. Chiu, H. S., Somvanshi, S., Patel, E., Chen, T. W., Singh, V. P., Zorman, B., ... & Lai, P. H. (2018). Pan-cancer analysis of lncRNA regulation supports their targeting of cancer genes in each tumor context. *Cell Reports, 23*(1), 297–312.
7. Clarke, R. (1994). Roger Clarke's Human Id in Info. Systems 11/11/03, 8: 59 PM Human Identification in Information Systems: Management Challenges and Public Policy Issues. *Information Technology & People, 7*, 6–37.
8. Dantcheva, A., Dugelay, J. L., & Elia, P. (2010, October). Person recognition using a bag of facial soft biometrics (BoFSB). In *2010 IEEE International Workshop on Multimedia Signal Processing, 3*(2), 511–516.
9. Dantcheva, A., Elia, P., & Ross, A. (2015). What else does your biometric data reveal? A survey on soft biometrics. *IEEE Transactions on Information Forensics and Security, 11*(3), 441–467.
10. Denman, S., Fookes, C., Bialkowski, A., & Sridharan, S. (2009, December). Soft-biometrics: unconstrained authentication in a surveillance environment. In *2009 Digital Image Computing: Techniques and Applications, 4*(1), 196–203.

11. Denman, S., Kleinschmidt, T., Ryan, D., Barnes, P., Sridharan, S., & Fookes, C. (2015). Automatic surveillance in transportation hubs: No longer just about catching the bad guy. *Expert Systems with Applications, 42*(24), 9449–9467.
12. Frischholz, R. W., & Dieckmann, U. (2000). BiolD: a multimodal biometric identification system. *Computer, 33*(2), 64–68.
13. Ho, T. K., Hull, J. J., & Srihari, S. N. (1994). Decision combination in multiple classifier systems. *IEEE Transactions on Pattern Analysis and Machine Intelligence, 16(*1), 66–75.
14. Jain, A. K., Dass, S. C., & Nandakumar, K. (2004, July). Soft biometric traits for personal recognition systems. In *International Conference on Biometric Authentication, 3*(1), 731–738.
15. Jain, A. K., Ross, A., & Prabhakar, S. (2004). An introduction to biometric recognition. *IEEE Transactions on Circuits and Systems for Video Technology, 14*(1), 4–20.
16. Jain, A., & Verma, C. K. (2012). A framework based on hybrid biometrics for personal verification systems. *International Journal of Applied, 1*(1), 55–58.
17. Jain, A., Nandakumar, K., & Ross, A. (2005). Score normalization in multimodal biometric systems. *Pattern Recognition, 38*(12), 2270–2285.
18. Jaramillo, M. C., & Zhang, D. D. (2013). The emerging role of the Nrf2–Keap1 signaling pathway in cancer. *Genes & Development, 27*(20), 2179–2191.
19. Khalifa, A. B., & BenAmara, N. E. (2012). Contribution to the fusion of biometric modalities by the choquet integral. *International Journal of Image, Graphics and Signal Processing, 4*(10), 1.
20. Koppen, M., Franke, K., & Vicente-Garcia, R. (2006). Tiny GAs for image processing applications. *IEEE Computational Intelligence Magazine, 1*(2), 17–26.
21. Lam, L., & Suen, S. Y. (1997). Application of majority voting to pattern recognition: an analysis of its behavior and performance. *IEEE Transactions on Systems, Man, and Cybernetics-Part A: Systems and Humans, 27*(5), 553–568.
22. Lee, J. E., Tong, W., Jin, R., & Jain, A. K. (2011). Image retrieval in forensics: application to tattoo image database. *IEEE Multimed, 10*, 1–20.

23. Min, R., Hadid, A., & Dugelay, J. L. (2011, March). Improving the recognition of faces occluded by facial accessories. In *2011 IEEE International Conference on Automatic Face & Gesture Recognition (FG), 42*, 442–447.
24. Nandakumar, K., & Jain, A. K. (2015). Biometric template protection: Bridging the performance gap between theory and practice. *IEEE Signal Processing Magazine, 32*(5), 88–100.
25. Niinuma, K., Park, U., & Jain, A. K. (2010). Soft biometric traits for continuous user authentication. *IEEE Transactions on Information Forensics and Security, 5*(4), 771–780.
26. Noor, S. W. N. A. M., & Ali, J. (2009). Technology trust and e-banking adoption: The mediating effect of customer relationship management performance. *The Asian Journal of Technology Management, 2*(2), 1–10.
27. Park, U., & Jain, A. K. (2010). Face matching and retrieval using soft biometrics. *IEEE Transactions on Information Forensics and Security, 5*(3), 406–415.
28. Park, U., Tong, Y., & Jain, A. K. (2010). Age-invariant face recognition. *IEEE Transactions on Pattern Analysis and Machine Intelligence, 32*(5), 947–954.
29. Ross, A., & Poh, N. (2009). Multibiometric systems: Overview, case studies, and open issues. *Handbook of Remote Biometrics, 2*(1), 273–292.
30. Saini, N., & Sinha, A. (2011). Soft biometrics in conjunction with optics based biohashing. *Optics communications, 284*(3), 756–763.
31. SANDERSON, C., & Paliwal, K. K. (2002). Information fusion and person verification using speech and face information. *Research Paper IDIAP-RR, 12*, 02–33.
32. Shen, W., & Tan, T. (1999). Automated biometrics-based personal identification. *Proceedings of the National Academy of Sciences, 96*(20), 11065–11066.
33. Siff, A. (2017). New MTA Towers Can Read License Plates, and Maybe More. NBC (New York). September, 28, 2–10.
34. Srinivasa, K. G., & Gosukonda, S. (2014). Continuous multimodal user authentication: coupling hard and soft biometrics with support vector machines to attenuate noise. *CSI Transactions on ICT, 2*(2), 129–140.

35. Tiwari, S., Singh, A., & Singh, S. K. (2012). Integrating faces and soft-biometrics for newborn recognition. *Int. J. Adv. Comput. Eng. Archit, 2*(2), 201–209.

36. Tome, P., Fierrez, J., Vera-Rodriguez, R., & Nixon, M. S. (2014). Soft biometrics and their application in person recognition at a distance. *IEEE Transactions on Information Forensics and Security, 9*(3), 464–475.

37. Ulery, B., Hicklin, A., Watson, C., Fellner, W., & Hallinan, P. (2006). Studies of biometric fusion. *NIST Interagency Report, 2*, 7346.

38. Velardo, C., & Dugelay, J. L. (2012, May). Improving identification by pruning: A case study on face recognition and body soft biometric. In *2012 13th International Workshop on Image Analysis for Multimedia Interactive Services, 4*(1), 1–4.

39. Zadeh, L. A. (1996). Fuzzy sets. In *Fuzzy Sets, Fuzzy Logic, and Fuzzy Systems: Selected Papers by Lotfi A Zadeh, 353*, 394–432.

40. Zewail, R., Elsafi, A., Saeb, M., & Hamdy, N. (2004, July). Soft and hard biometrics fusion for improved identity verification. In *The 2004 47th Midwest Symposium on Circuits and Systems, 1*, 1–225.

41. Zhang, Y. K., Li, X., Zhao, H. R., Jiang, F., Wang, Z. H., & Wu, W. X. (2019). Antibodies specific to membrane proteins are effective in complement-mediated killing of *Mycoplasma bovis. Infection and Immunity, 87*(12), 40–119.

Chapter 4

Eye Recognition Technique and its Characteristics

CONTENTS

4.1. INTRODUCTION

Our eyes are special, similar to the other parts of our body, and can be employed for biometric purposes. In our eyes, two basic elements have great biometric entropy. The eye's iris is the first, and the second is the eye's retina, both of which are not visible to the human eye. On these two biometric traits, the industry of biometric recognition is based (Pan & Xie, 2005). In 1994, the first blatant automatic iris recognition was issued.

The retina and iris, as internal components of the eye, are extremely resistant to harm. Although the iris color and shape are innately dependent, each person's retina and iris patterns are unique to them (this also put on to monozygotic twins) (Mallat, 1989):

a) The cornea is the clear covering that covers the front of the eye. It is a translucent connective tissue that permits light to enter the eye in conjunction with the lens. Its poor curvature causes astigmatism.
b) With intraocular fluid, the front chamber is filled that is replenished regularly.
c) The iris is a circularly structured musculature that enlarges/narrows the pupil and shapes an annulus.
d) The pupil is a small aperture in the middle of the iris that controls how much light enters the eye.
e) On the ciliary body, the lens is suspended and can bend, causing the refractive index to change. The eye will be unable to accommodate if the lens loses this ability (focus).
f) The sclera is a transparent white film that covers the entire eyeball and penetrates through the cornea in the front.
g) The inside of the mesh is filled with vitreous fluid.
h) The retina is the inner layer of the eye that contains light-sensitive cells. It displays the image in the same way as a camera does.
i) A vast number of nerve fibers go in the central nervous system through the optic nerve (CNS) (Mallat, 1989).

Biometrics and ophthalmology are the two scientific fields that cope with eye features. Ophthalmology is a branch of medicine that studies and treats the eye's health and its surrounding areas. Iridology is mentioned to be thorough.

The special characteristics of the eye do not change with time, and they are so distinct that in the field of biometrics, it is possible to unambiguously identify two distinct individuals away from each other to validate that person's identity.

4.2. RECOGNITION THROUGH IRIS

At a glimpse, the colored region of the eye is the iris that we can perceive in other individuals. The amount of light that goes in the eye is regulated through the iris, similar to how a camera aperture regulates the amount of light passing through the lens. The pupil is the black hole in the middle of the iris. Fine muscles that inflate or narrow the iris are related to the iris. Each person's iris is unique in color, texture, and pattern, similar to how fingerprints are unique. The chances of finding two identical irises, on the other hand, are substantially lower than with fingerprints (Drahanský & Yang, 2018). The clamping muscle runs sideways the iris's edge and pushes the eye inwards into a brighter light. When the light is dim, the stretching muscle transits transversely, like stretching the iris with a bicycle string. The iris is a flat membrane that separates the front and back regions of the eye. The structure of the human eye and the locations of its portions are depicted in Figure 4.1. Melanin, a pigment, is responsible for the iris' color. It is situated between the sclera and the pupil of the eye. The iris is around 11 mm in diameter. Its visual texture appears in the 3rd month of pregnancy and develops over two years (Kronfeld, 1962). The underlying structure of the iris does not change during life, and the iris is unique even in twins. Figure 4.2 depicts the structure of the iris.

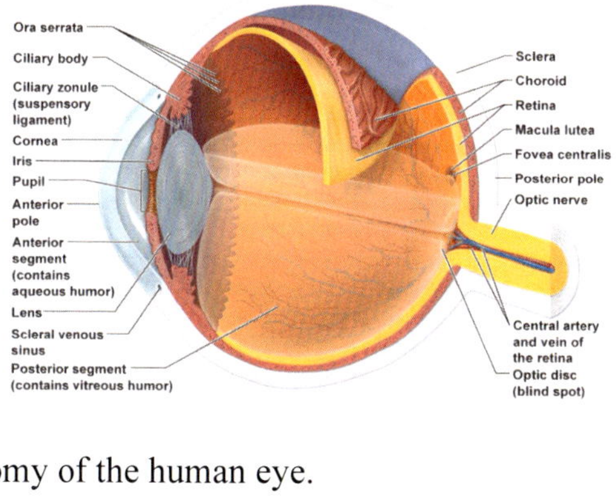

Figure 4.1. Anatomy of the human eye.

Source: *https://www.allaboutvision.com/resources/anatomy.htm.*

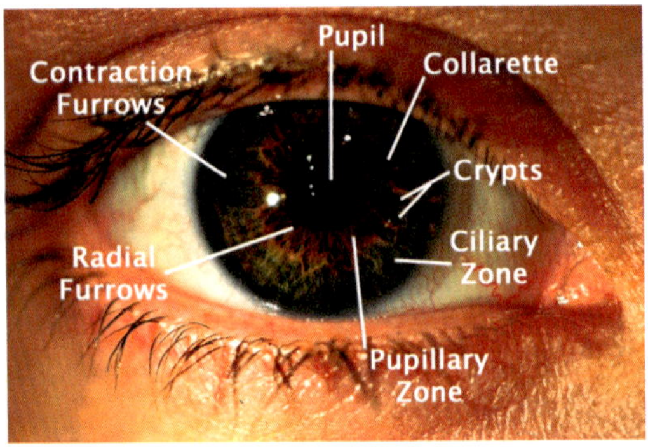

Figure 4.2. Structure of the iris—features

Source: *https://medlineplus.gov/ency/imagepages/8867.htm*

The iris' surface is highly intricate. John Daugman defined the 250 characteristics of the iris. The following are very significant for identification (Kronfeld, 1962):

a) *Crypts*: These are the thinnest parts of the iris, with decay in front and stroma making its distinctive drawing.

b) *Radial furrows*: Through the pupil to the collar, a chain of extremely fine razor-shaped nibs.

c) *Pigment spots*: Pigment clusters appear at random on the iris' surface.

4.2.1. Influence of Light on Iris Acquisition

In the spectrum of visible light, the light we see is an electromagnetic wave. The wavelength of each of these waves is different. Colors are seen as diverse wavelengths of the visible spectrum, nevertheless, the eye is also sensitive to further wavelengths (Roberts, 2005):

a) 100–315 nm: Most of the substance is absorbed in the cornea, with the remainder disseminated in ventricular water.

b) 315–400 nm: captivated in the lens.

c) 400–1400 nm: On the retina, the light goes through the lens. The eye reacts in 0.25 seconds to visible light in the 400–700 nm range.

d) Moreover, 1400 nm: absorbs the cornea, producing severe tearing and raising the body's warmth.

The visible layers, particularly on the iris, can be seen in visible light. Because melanin absorbs visible light, it shows less textural information than IR (infrared light).

Infrared (IR) light melanin, on the other hand, is more user-friendly and is favored for iris detection since it does not irritate or create the negative feelings associated with eye lighting.

For iris recognition, there are four simple schemes:

a) *Gabor demodulation*: Each iris design is demodulated to acquire phase information for feature extraction (Daugman, 2009).

b) *Wavelet features*: Using the wavelet transform, extract the vector of features (Lim et al., 2001).

c) *Analysis of independent components*: As a vector of features, independent component analysis factors are employed (Jun et al., 2010).

d) *Local keys variation*: Wavelet transformation extracts features from a series of concentrations of one-dimensional signals to represent relevant information (Ma et al., 2004).

4.2.2. Gabor's Demodulation

The first stage of Daugman's algorithm or Gabor's demodulation is locating the iris in the captured image. The iris should be perused accurately for phase diagrams to be created, containing information around the orientation, position, and several unique identification markers. Afterward, the template has been extracted, the database is searched for it. Figure 4.3 depicts Daugman's algorithm.

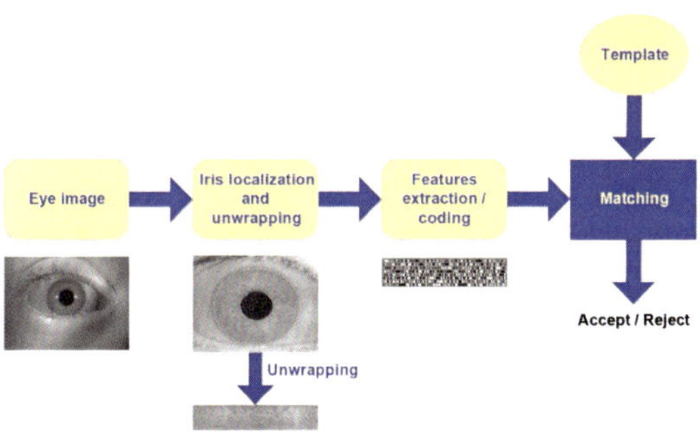

Figure 4.3. The identification process of Daugman's algorithm.

Source: *https://www.intechopen.com/chapters/60581*

In the image of the eye, and iris (curve boundary) is first positioned. With the subsequent operative, the iris may be found.

$$\max_{(r,x_0,y_0)}\left|G_\sigma(r) * \frac{\partial}{\partial r}\oint_{r,x_0,y_0}\frac{I(x,y)}{2\pi r}ds\right| \quad (1)$$

$G_\sigma(r)$ is the Gaussian smoothing function, σ, $I(x,y)$ is the raw input picture, and the operative looks for the maximum in the blurred partial derivative of the image respecting the radius r and the center coordinates (x_0, y_0). The operator is a circular edge detector, and it returns the highest value if the candidate circle has the same pupil center and radius as the pupil. Figure 4.4 depicts examples of localized irises (Norn, 1985).

Locating the lid is the succeeding step. The approach used to identify the upper and lower eyelids is the same as that used to determine the position of the iris. The component of the preceding formula (Eq. (1)) used to detect the contour is exchanged by a circular arc, with the parameters optimized to match each eyelid boundary using normal statistical estimate methods. Figure 4.5 depicts an example of confined lids (Brown & Wald, 1964).

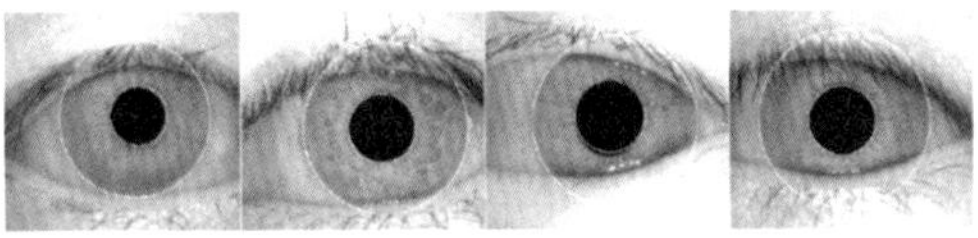

Figure 4.4. Examples of localized irises.

Source: *https://scialert.net/fulltext/?doi=itj.2008.924.929*

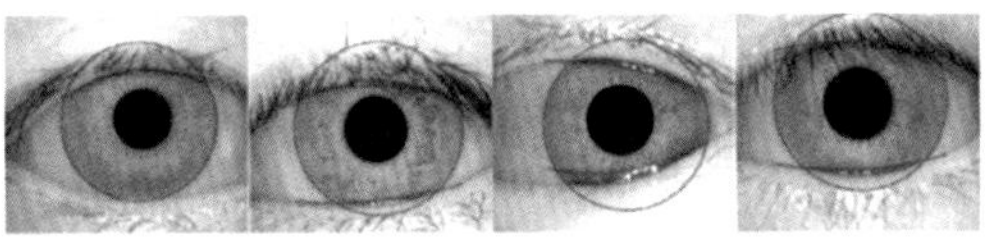

Figure 4.5. Examples of localized lids.

Source: *https://www.intechopen.com/books/6563*

4.2.3. Daugman's Gross Alignment Model

Every point inside the iris to polar coordinates(r, θ) via Daugman's gross alignment model, where r is from the interval $\langle 0, 1\rangle$ and θ is the angle from the interval $\langle 0, 2n\rangle$. The model pays for pupil dilation and expansion caused by the size and translation invariance of the polar coordinate system. The model, however, does not account aimed at rotational variation, which is remedied by rotating the iris pattern in the path of the during the contrast stage till both templates are identical. Figure 4.6 depicts the introduction to the coordinate system (Hangai et al., 2006).

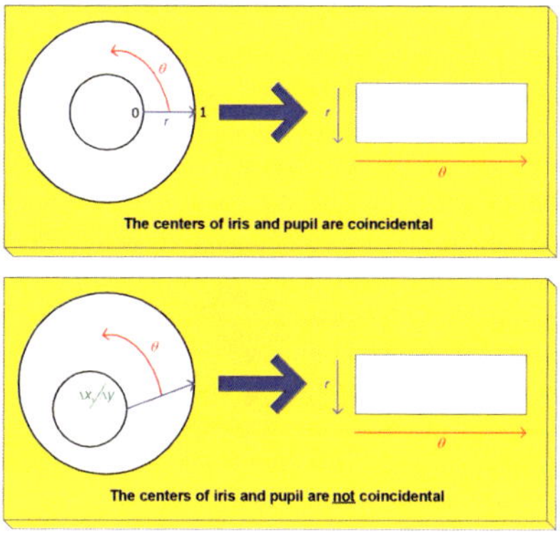

Figure 4.6. Implementation of Daugman's algorithm coordinate system.

Source: *https://www.intechopen.com/chapters/60581.*

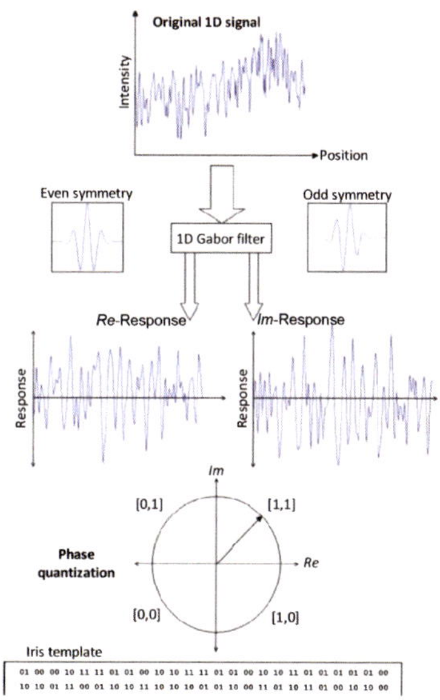

Figure 4.7. Illustration of the encoding process.

Source: *https://www.researchgate.net/figure/An-illustration-of-the-feature-encoding-process_fig4_266160074.*

4.2.4. Iris Features Encoding

The iris code is 2048 bits long (256 bytes). The input image is 64 × 256 bytes in size, the iris code is 8 × 32 bytes in size, and the Gabor filter is 8 × 8 bytes (Figure 4.7). Figure 4.8 shows an example of the iris code.

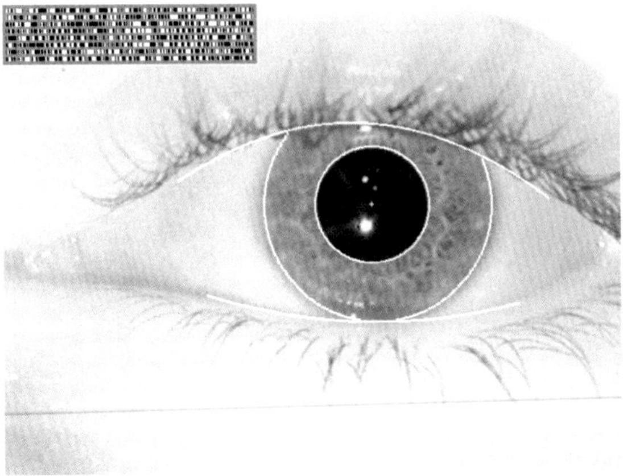

Figure 4.8. Example of an iris code.

Source: https://www.researchgate.net/figure/Example-of-an-iris-image-segmented-using-Daugmans-algorithm-with-the-IrisCode-bit_fig1_244403267.

4.2.5. Comparison of Iris Codes

The Hamming distance among the two 256 housing codes is used to make the comparison. The total of exclusive totals among bits is employed to compute the Hamming distance of two iris codes, A and B.

$$HD = \frac{1}{N}\sum_{j=1}^{N} A_j \otimes B_j \qquad (2)$$

where $N = 2048$ (8 × 256), except the iris is sheltered through the lid. Then, just effective areas are employed to compute the Hamming distance.

When both samples come from a similar iris, the Hamming distance is zero or near zero. One design is shifted to the left/right to ensure rotational consistency, and the equivalent Hamming distance is always planned. The comparison score is then calculated using the Hamming distance's lowest value. Figure 4.9 shows an example of how to liken iris codes employing shifts (Stanetić et al., 2016).

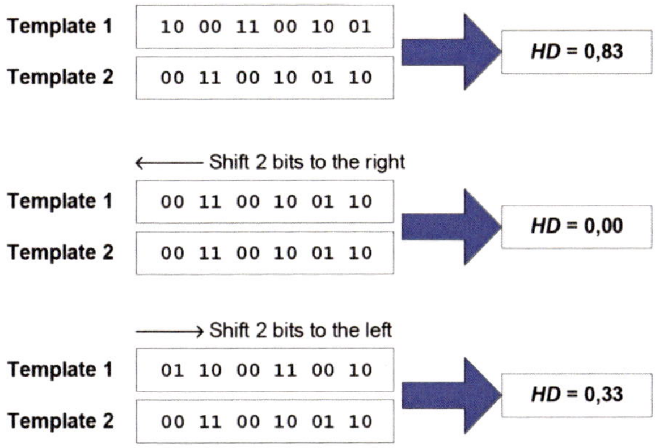

Figure 4.9. Example of the comparison of iris codes using shifts.

Source: *https://www.intechopen.com/books/6563.*

4.2.6. The Disadvantages and Advantages of the Iris for Biometric Identification

For biometric identification systems, there are few advantages of employing an iris that is given below (Scanlon et al., 2009):

a) During an individual's life, the iris is constant.

b) Snapshots are user-friendly and noninvasive.

c) The template is tiny in size.

d) The iris is a reasonably well-protected internal organ from external effects.

e) The iris contains far more biometric entropy data than fingerprints.

The following are some of the drawbacks of utilizing the iris for identification (Timberlake & Kennedy, 2005):

a) The absence of a method to protect against iris spoofing or contact lens spoofing.

b) The obstruction could also be a result of consumers' fears that the scanner will harm their eyes.

The limits of iris recognition are summarized in the following list. In some ways, these could be counted among the drawbacks (Simon, 1935).

a) The attainment of an iris image necessitates user participation; the user must be in front of the camera at a predetermined position and distance. Although certain technologies allow for semi-automated to instinctive scanning, the rate of error of these systems is motionless very significant.

b) For high-performance systems, their cost is relatively high.

c) Iris images may be of poor quality, leading to verification, registration, or identification mistakes.

d) The iris might alter over time, especially if you are suffering from a variety of ailments. Cataract surgery and disorders like nystagmus or aniridia can cause the iris to change. Due to the clouding of the eyes, the iris may not be visible at all for certain blind persons.

e) Individual sections of the iris are linked to several interior tissues of the human body, posing the risk of misinterpreting the scanned design to evaluate a person's health. Iridology is a branch of alternative medicine (Ma et al., 2004).

4.3. RECOGNITION BY RETINA

Added option provided through the eye is retinal recognition. Obtaining a great-quality eye image is maybe the most difficult component of the entire retinal identification technique. The concepts of medical instruments for the inspection of an eye can be found here. It is also vital to comprehend the retina's role in human visualization and its position and the materials limited within to perform biometric identification (Durlu, 2021).

4.3.1. Anatomy of the Retina

The retina is taken as a portion of the CNS (central nervous system). This is the lone noninvasively visual component of the CNS. With a thickness of 0.2–0.4 mm, it is a light-sensitive layer of cells positioned near the rear of the eye. It detects light rays that enter the eye over an eye lens and the pupil that rotates and reverses the image. The retina is a complicated structure of multiple layers of neurons connected through synapses (Figure 4.10). Photoreceptors are the only neurons that respond to light directly. Cones and rods are the two basic forms of these. The retina comprises 72% of the internal eye in humans (Elliott et al., 2004). There are around 7 million cones and 75–150 million rods on the retina's total surface. The eye would be compared to a 157 MP camera in this case. Rods are utilized to sense light and provide black and white vision in response to the impact of one to two photons. Cones are employed to sense colors and classified into three kinds based on the base color they are delicate to (green, red, or blue). However, they are less subtle to light intensity. In these cells, a process known as transduction occurs, in which a series of electrical and chemical events are converted into electrical impulses. These are subsequently sent to the CNS via the optic nerve (Saraereh et al., 2020).

On the retina of an eye, we may see the two most distinct points. It's a macula and a blind spot. The optic nerve enters the eye at a location called a blind spot, which is around three mm^2 in size and devoid of any receptors.

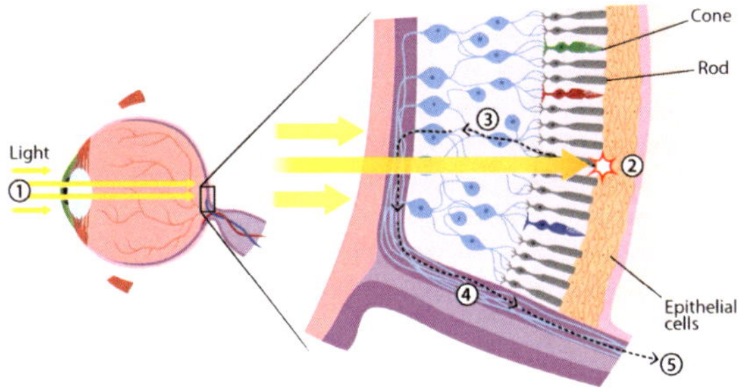

Figure 4.10. Structure of the retina.

Source: https://www.pinterest.com/pin/306737424614797595/.

As a result, if the image goes down into the blind spot, it will be hidden from view. In order to fill this space, the brain frequently "guesses" how the image should appear. Figure 4.11 illustrates how to check for the presence of a blind spot. When we look at the cross with our left eye closed, the black circle fades away at a specific distance from the image. This is the precise point at which the image settles on a blind area (Roberts, 2005).

In contrast, the macula is the clearest visual area, with a diameter of around 5 mm and a predominance of cones. This part contains the highest number of light-sensitive cells, decreasing as you get closer to the margins. The fovea, which describes visual acuity and receptor concentration, is located in the macula's center. This area reflects our point of view. Surprisingly, the macula (yellow spot) is slightly redder than the nearby area rather than yellow. On the other hand, this trait was conferred by the fact that yellow arises after a person's death (Vander et al., 2001).

The choroid, which is a layer that lies among the sclera and the retina, feeds the retina. It has blood vessels and a pigment that absorbs light. The retina is intricately intertwined with healthy arteries and nerves, as shown in Figure 4.12. It depicts a brain-like apparatus in which the structure and venous tangle stay constant through life. The retina receives blood from two key sources: the retinal vessels and the artery. The blood artery that supplies the retina's outer layer of photoreceptors receives more blood flow. The retinal artery, which predominantly feeds the interior of the retina, provides another blood supply. There are normally four primary branches of this artery (Saraereh et al., 2020).

External impacts are highly protected from the retina, which is positioned inside the eye. The vessel pattern does not vary during life, making it ideal for biometric reasons.

Figure 4.11. Blindspot testing.

Source: *https://visionaryeyecare.wordpress.com/2008/08/04/eye-test-find-your-blind-spot-in-each-eye/.*

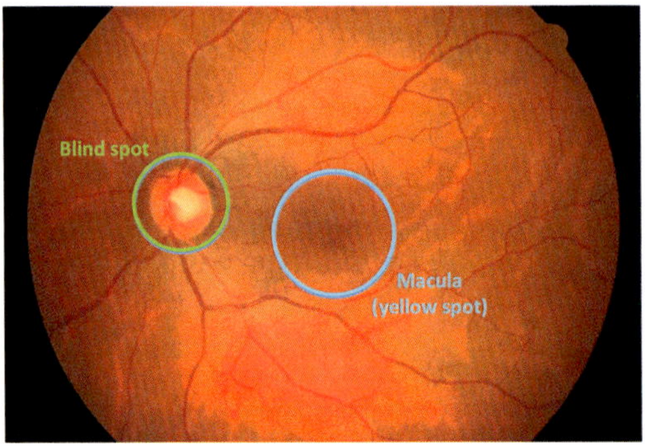

Figure 4.12. The fundus camera took a snapshot of the retina.

Source: *https://link.springer.com/chapter/10.1007/978–3-030–27731–4_11.*

The retina acquires a camera-like image. In the same way that film appears in the lens's focus on the retina, the beam traveling over the pupil looks in the lens's focus on the retina. In medical practice, specialized optical equipment is utilized to examine the retina visually (Mele & Federici, 2012).

4.3.2. Eye Diseases

The iris is not particularly exciting in ophthalmology, when we ignore the thrilling and extremely occasional cases of a disease, pigment changes frequently happen, which are not the outcome of a disease and have no impact on human health. In ophthalmology, the core focus is inspecting the eye's retina while also considering the eye's overall health. There are several disorders and damage to the retina that medical specialists are interested in, nevertheless, they are all covered in an encyclopedia of ophthalmology along with several pages (Doherty et al., 2010). Diabetes and age-related macular deterioration (ARMD) make up the majority of the cases. Hemorrhages or druses can occasionally form in the retina; nevertheless, as previously stated, potential injury or retinal disease should be addressed by ophthalmologists. We process video or photo sequences to look for diseased signs in the meantime our research group works with medical experts. At the moment, we are concentrating on delimiting and detecting hemorrhages and exudates in images and automatically detecting the blind spot and macula. These are the landmarks that we use to pinpoint the site of abnormal findings. The fovea centralis, which is where the sharpest vision is found, is the worst section. When this area is destroyed, it has a major influence on our vision. Figure 4.13 illustrates one method of detecting abnormal symptoms. In the retina, by monitoring the eminence of blood flow, we can also deal with coworkers. Because the input data is so diverse, there is still much work to be done in the entire area of medical video and imaging processing. The finest diagnostic tool, for the time being, is a medical practitioner (Hornof et al., 2003).

Sickness can affect any portion of the human body, whether it is treatable or not. An inveterate condition will be defined as an impairment that cannot be medically or else removed deprived of erasing biometric data (for example, elimination). The curable condition is easily treatable and has few side effects. Both of these disorders can cause damage to the retina. These illnesses can have a big impact on how long it takes to recognize someone. If a disease alters the retina's structure, the pattern may be evaluated incorrectly or completely rejected (Dobeš et al., 2004).

4.3.2.1. Macular Degeneration

Macular degeneration, also known as age-related macular degeneration, is a disease that affects 90% of people as they get older (ARMD). In the remaining 3%, macular degeneration manifests itself in the type of Best's macular deterioration or Stargardt's disease in children or teenagers. These disorders are passed on via the generations (Trokielewicz et al., 2014).

The region of the retina that makes the middle of the field of vision is desecrated in macular degeneration (Figure 4.14). Thus, there is a significant disturbance in the center field of vision. The patient notices a grey shadow in the center that descends to a black spot. The macula's peripheral visualization is unaffected.

Figure 4.13. Highlighted hemorrhage (right), detection of suspected areas (center), and Hemorrhage (left).

Source: *https://www.intechopen.com/chapters/60581.*

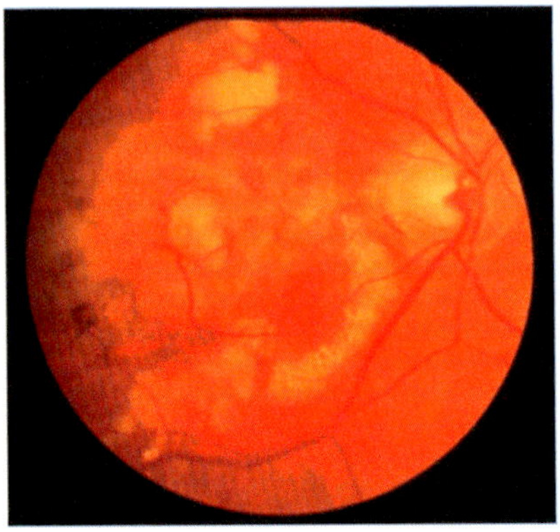

Figure 4.14. Macular degeneration.

Source: *https://en.wikipedia.org/wiki/Macular_degeneration.*

There are two types of macular degeneration: dry (atrophic) and wet (proliferative) (exudative). A fuzzy grey or black patch in the center of the field of vision is one of the most typical symptoms. Deformed straight lines, fuzzy letters, or unsuitable shapes of various items are visible to the affected person. It also appears to be affecting color vision, which has diminished. One or both eyes' side vision stays sharp (Stanetić et al., 2016).

4.3.2.2. Diabetic Retinopathy

Diabetic retinopathy (DR) is a retinal stirring condition. It develops as a result of diabetes mellitus overall blood vessel damage. Diabetes misdiagnosed causes small catheters in the eyes to clog, producing blood flow to stall. The retina is also impacted when the vessels "leak," allowing fluid to escape and causing the retina to enlarge. Inadequate blood circulation and retinal edema impair vision. The eye attempts to correct the problem through forming novel blood vessels, but these are deprived and damaging, cracking, causing hemophthalmos, and causing retinal traction detachment. There are two types of diabetic retinopathy: proliferative and non-proliferative (Scanlon et al., 2009).

4.3.3. Toxoplasmosis

Toxoplasmosis is a zoonotic illness, meaning it can be transmitted from animals to people. It can be found around the world. Anti-toxoplasmosis antibodies are formed by 10–60% of European countries, dependent on dietary choices. Seropositivity is about 20–40% in the Czech Republic. Higher temperatures, flu-like symptoms, weariness, headaches, and swollen lymph bulges are the most common symptoms of the disease.

An acute infection can progress to a chronic stage, although it typically goes unreported and is just detected through the presence of special anti-toxoplasmic antibodies in the blood, which can persevere at low levels throughout their lives. Nodal, ophthalmic (see Figure 4.16), cerebral, and gynecological illnesses are among the many types of illness. Toxoplasmosis in other forms is uncommon (Drahansky & Yang, 2018).

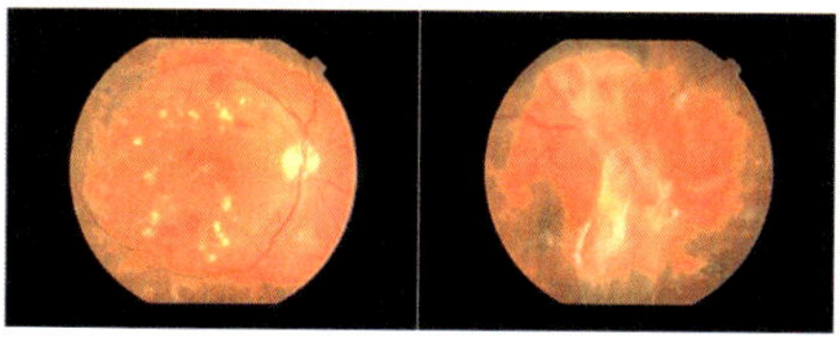

Figure 4.15. Non-proliferative and proliferative diabetic retinopathy.

Source: *https://www.researchgate.net/figure/Retinal-image-with-some-DR-features-Mild-non-proliferative-diabetic-retinopathy-left_fig7_338646380.*

4.3.4. Retinal Examination Tools

A direct ophthalmoscope is the most popular equipment for studying the retina. The patient's eye is inspected through the pupil with an ophthalmoscope at a distance of several millimeters.

A source of light from a mirror or a semipermeable mirror with a hole at a 45° angle in the observation axis illuminates the retina (Timberlake & Kennedy, 2005). There are numerous kinds of ophthalmoscopes currently available. The principle, however, is fundamentally similar: the investigator's and the investigated's eyes are on the same axis, and the retina is lighted by a light source from a mirror or a semipermeable mirror with a hole in the observation. A direct ophthalmoscope's disadvantages include a small investigation area, the need for experience in handling, and patient participation. The so-called fundus camera, which is now most probable to have the utmost standing in inspecting the retina, is employed for a more detailed inspection of the ocular background. It enables color photography to cover nearly the entire retinal surface, as seen in Figure 4.12. This device's visual concept is dependent on so-called indirect ophthalmoscopy (Timberlake & Kennedy, 2005). A white light source illuminates the retina, which is subsequently scanned with a CCD (charge-coupled device) sensor in fundus cameras. Some types can also use frequency analysis of the scanned image to determine the retina's center and focus it automatically.

4.3.5. Histology of Retinal Recognition

Isidore Goldstein and Carleton Simon, ophthalmologists, found eye illnesses in which the bloodstream image in two persons in the retina was distinct for every individual in 1935. They then published a scientific paper on utilizing the retinal vein picture as an exceptional design for identification (Simon, 1935). Dr. Paul Tower, who printed an essay on examining monozygotic twins in 1955, backed their research.

He found that the retinal vascular patterns have the least in common with the other patterns he looked at. Identification of the vessel's retina was an everlasting notion at the time.

in 1975 Robert Hill, who founded EyeDentify, dedicated nearly all of his effort and time to develop a simple, totally automatic equipment able to extracting a photo of the retina and authenticating the user's identification. Functional devices, on the other hand, did not seem on the market for numerous years.

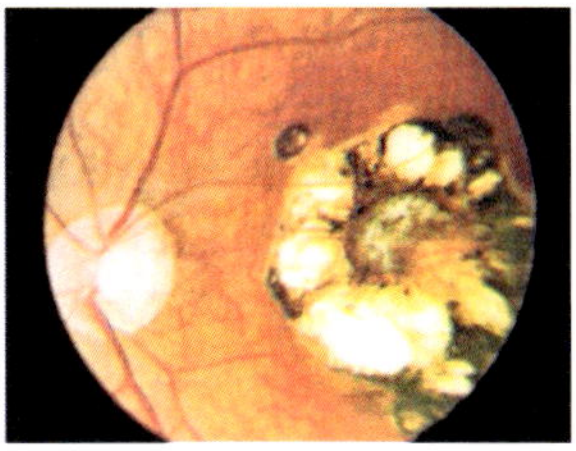

Figure 4.16. An eye affected by toxoplasmosis.

Source: *https://emedicine.medscape.com/article/2044905-overview.*

Numerous other firms sought to alter existing fundus cameras to retrieve the retina's image for identifying purposes. Nevertheless, there were several notable drawbacks to these fundus cameras, including the rather difficult arrangement of the visible light spectra, optical axis, which made identification very rough for users, and, finally, the exorbitant charge of these cameras (Tower, 1955).

More research guide the use of IR (infrared) illumination, which is practically transparent to the choroid coat, which reflects the radiation to generate an image of the blood vessels in the eye. Because IR illumination is invisible to humans, the pupil diameter does not shrink when the eye is irradiated.

The device's first functional prototype was produced in 1981. An ordinary personal computer was connected to the apparatus with an eye-optic camera to light the infrared radiation for image capture and processing. A basic correlation comparison approach was identified as the most acceptable after considerable testing (Metzner et al., 2009).

EyeDentify Inc. has released the Eye Dentification System 7.5, which performs verification dependent on the PIN given by the user and on the retina image with the data contained in the database, after another four years of hard work.

The apparatus took a circular photograph of the retina. The image was reduced from 256 twelve-bit logarithmic models to a reference record of 40 bytes for every eye. The temporal domain is where contrast information is kept. Furthermore, in the time domain, 32 bytes were added for each eye to speed up recognition (Hill, 1996).

4.3.6. Technology and Principles

The device's functional concept can be broken down into three non-trivial subsystems. (Hill, 1996):

a) *Image, signal attainment, and dispensation*: The camera and optical system should be proficient in taking a numerical image of the retina that can be processed.

b) *Comparison*: a program that extracts essential elements through a perused image and likens them to a database of designs on a device or computer.

c) *Representation*: Every retinal image should be signified in a form that can be swiftly associated and saved in a database.

4.3.6.1. *Sensing optical system*

We will look at how to employ sensing devices to take photographs of the back or front of the eye. Indirect and direct ophthalmoscopy, along with the most generally employed inspection, a slit lamp, which allows for biomicroscopy of the anterior segment of the eye, are the principal ophthalmoscopic inspection procedures for the posterior and anterior sections of the eye. A fundus camera, also known as a retinal camera, is special equipment that allows you to see the posterior segment of the peripheral section, yellow spots, and optic nerve of the retina (see Figure 4.17 on the right) (Foster et al., 2010). It operates on the idea of indirect ophthalmoscopy, with a primarily white light source integrated into the equipment. Numerous filters can alter the light, and the optical system is engrossed in the patient's eye, reflecting off the retina and returning to the fundus camera lens. There are two types of mydriasis: non-mydriatic and mydriatic. The difference is whether or not the patient's eye should be put into mydriasis.

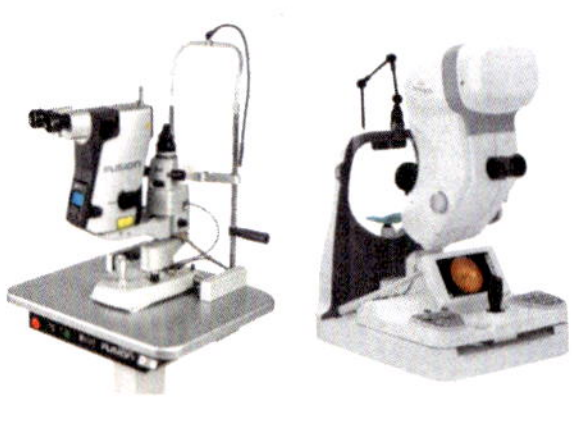

Figure 4.17. At the right, there is a non-mydriatic fundus camera while on the left there is a slit-lamp.

Source: *https://sunnymed.en.made-in-china.com/product/ydMQPCXOkUVx/China-Sy- V0 36A-Best-Price-Ophthalmic-Non-Mydriatic-Digital-Eye-Fundus-Camera.html.*

The goal of mydriasis is to dilate the pupil of the human eye to the "inlet opening" which is greater, permitting one to read a bigger portion of the retina. The non-mydriatic fundus cameras are favored since the patient can leave directly after the inspection and drive a car, which is not feasible with mydriasis.

Mydriasis is, nevertheless, required in some cases. The cost of these medical gadgets is in the tens of thousands of Euros, with just specialized medical offices determining the price (Elliott et al., 2004).

The optical device's mechanical construction is quite complicated. The scanning equipment works on the same premise as medical eye-optic instruments. These so-called retinoscopes, also known as fundus cameras, are sophisticated equipment with a high price tag.

The idea is similar to a retinoscope: a beam of light is engrossed on the retina, and the reflected light is scanned through a CCD camera. The retinoscope's light beam is accustomed such that the eye lens emphasizes the retina's surface.

This imitates a part of the spread light beam back to the ophthalmic lens, which settles it so that the beam exits the eye at a similar angle as it entered (return reflection). As illustrated in Figure 4.18, a picture of the eye's surface can be acquired by rotating the camera along the visual axis by about 10 degrees. Because of the reflection of light through the cornea, which would be worthless throughout raster scanning, the gadget took a circular image of the retina.

EyeDentify's original products featured a sophisticated optical structure with revolving mirrors to cover the retinal area—this method is defined in U.S (Vander et al., 2001).

UV-IR cut filters are employed in the design to line up the scan axis with the visual axis. Figure 4.19 depicts a schematic illustration of the patent. From the camera, the distance between the lens and the eye was around 2–3 cm. The instrument's optical axis alignment system is a critical component, and it is detailed in greater detail in U.S (Foster et al., 2010).

EyeDentify's novel optical systems are significantly easier to operate and have the advantage of setting optical axes along with less user effort than earlier systems. A spinning scanning disc with multifocal Fresnel lenses is the most important component. The U.S. describes this construction (Daugman, 1993).

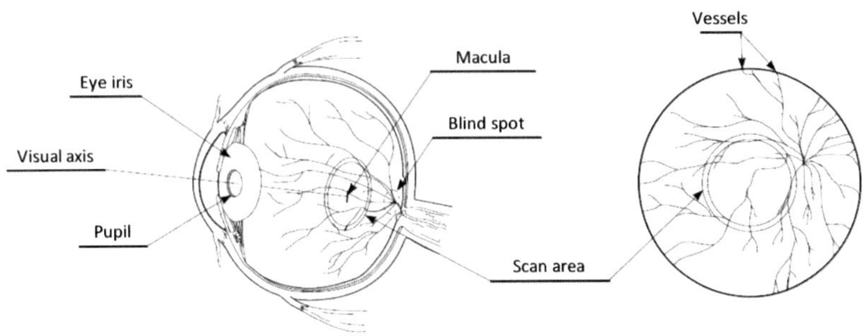

Figure 4.18. Efficient principle for finding a retinal image of the eye background.

Source: https://www.bookdepository.com/Machine-Learning-Biometrics-Jucheng-Yang/9781789235906.

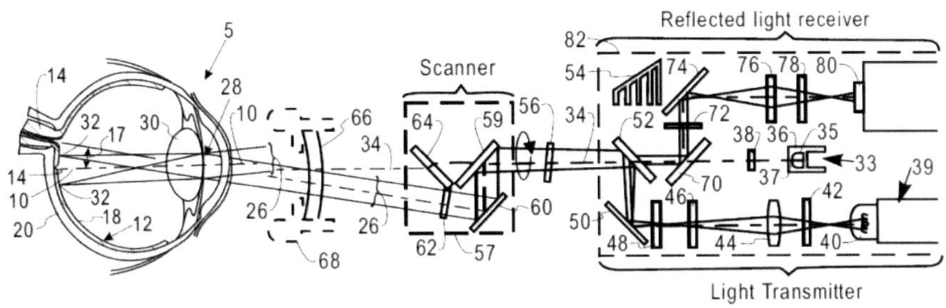

Figure 4.19. The first version of Eye Dentification System 7.5 optical system.

Source: *https://www.bookdepository.com/Machine-Learning-Biometrics-Juch eng- Yang/9781789235906.*

The target should be located in a similar place during the scanning period to guarantee that the region is absorbed on the retina and that the user's eye is in the axis of the scanning beam.

A variety of optical networks and focal lengths of 7, 3, 0, and +3 diopters can be used. It is believed that most users, regardless of their optical impairments, will be able to focus.

When the eye concentrates on a target, the device automatically aligns itself to the axis by centering the rotating disc to the eye background. When two or more optical designs are aligned overdue each other, the IR beam is positioned at the user's pupil, allowing the information to be delivered (Saraereh et al., 2020).

4.3.6.2. Comparison

When a person stares inside the camera's optical structure, their head may rotate somewhat away from the original scanned place. The data can be rotated by multiple degrees using the rotational algorithm. This method is repeated multiple times until the top match or the highest relationship is found.

The contrast of the attained samples is confirmed in various steps (Richardson & Spivey, 2004):

- The visual reference is turned into a field with a similar amount of elements as the attained field using sampling, ensuring arrangement.
- The intensity is normalized in each field so that RMS equals 1.
- A Fourier transform equal time sphere is used to correlate the field.

When the temporal shift is zero, the correlation value determines the comparator quality. It ranges from +1 (absolute match) to 1 (nearest match). A score of roughly 0.7 is a match in the past.

4.3.6.3. Representation

The retinal depiction is resultant from an annular region-based frame. The scanned area is chosen to accommodate the worst scanning situations, yet it is also large enough for biometric identification. It is not required to get an image with excessive resolution or area for these reasons (Jacob, 1993).

The scanned area is chosen to accommodate the worst scanning conditions, yet it is also large enough for biometric identification. It is not required to obtain an image with excessive area or resolution for these reasons (Turk & Pentland, 1991).

There were two major representations of the retinal picture in combination with an EyeDentify device:

- There are 40 bytes in the original form. The imaginary and real spectrum management provided by the Fourier transform encode contrast information.
- There are 48 bytes in the new representation. There is no time-domain contrast info in this. The real benefit of time depiction is that it allows quicker and more effective processing while using less computing power.

The template of the retina has 96 fields of 4-bit difference numbers derived from 96 time-domain images of concentric circles or 96× 4 = 48 bytes. Regularizing for this layout—4 bits of intensive design— intensity can take values in the 8.7> time interval (Bulling et al., 2010).

When we say new research in the retina, the condition is quite straightforward since the algorithms are looking for bifurcations and crossings in the image that indicate the person's position. Figure 4.20 depicts an example. When a stronger clinical condition impacts the extraction and detection of bifurcations and crossings in the retina, recognition becomes difficult.

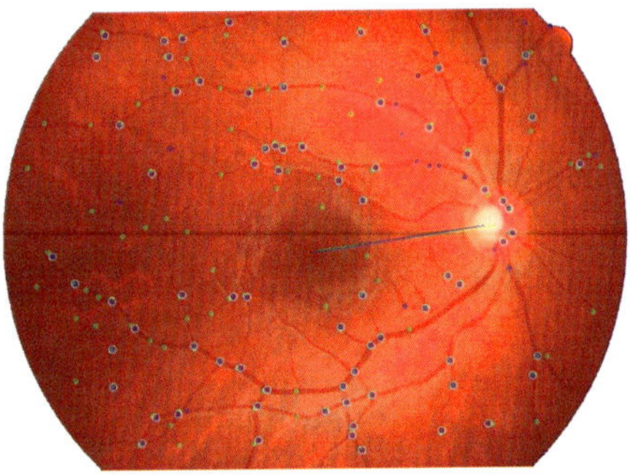

Figure 4.20. Extracted structures (crossings and bifurcations, including connection of blind spot and macula) in the retina.

Source: https://www.hindawi.com/journals/jhe/2020/7156408/.

Biometric systems contain information about personal health since a large amount of data can be read through the iris and retina. As a result, it is up to us to decide how much we will guard this personal data and whether or not we will employ the systems. Nevertheless, if the maker ensures that no health information is safeguarded against any potential security attacks, we may be delighted to utilize the system (Lykins et al., 2006).

4.3.7. Limitations

Retinal recognition may be the most limiting of mainstream biometrics. They are not definitive, nevertheless, there is now no system that can eliminate these flaws (Hill, 1996):

a) *Terror of eye damage*: Although the low level of infrared illumination utilized in this gadget is safe for the eyes, there is a common misconception among the general public that these devices can cause retinal impairment. To acquire confidence, entirely users must be aware of the system.

b) *Indoor and outdoor use*: The rate of erroneous rejection can be increased by having small pupils. Because the light must pass over the pupil twice, the return beam will be severely attenuated if the user's pupil is extremely small.

c) *Ergonomics*: The requirement to go close to the sensor may make the gadget less comfortable to use than other biometric approaches.

d) *Simple astigmatism*: People with astigmatism cannot emphasize their eyes on the point, preventing the template from being generated correctly.

e) *High price*: It is reasonable to expect that the gadget's price, particularly the retroviral optical device itself, will always be more than, say, voice or fingerprint recognition systems (Holsanova, 2014).

4.4. CHARACTERISTICS OF IRIS AND RETINA RECOGNITION TECHNOLOGY

The properties of retinal and iris recognition are discussed in the following subsection. Some of the traits are now apparent from the preceding subsections, which discussed the principles of processing and sensing these biometric features (Jacob, 1991).

4.4.1. Acceptance

4.4.1.1. Iris

Because there is no requirement for instant engagement with the user, iris identification has a moderate degree of acceptance. The user has to position ahead of the device and look at the sensor from a specified distance, deprived of twisting their head. The time it takes to capture and assess an image is roughly 2 seconds (Velisavljevic, 2009).

4.4.1.2. Retina

The acceptance rate for the retina is very poor. Many people are hesitant to use technology. They believe a laser will be utilized, which could cause damage to their eye. These fears, however, are unfounded since a laser is never employed in this scenario. Another issue is the technique for retrieving retinal images. This is tiresome, and some users may find it inconvenient (Clark et al., 1987).

Direct user engagement is also essential for the retina. At least with the existing technologies, the user needs to provide a significant amount of participation. As a result, acceptance is low.

4.4.2. Reliability

4.4.2.1. Iris

Because of ambient light, excessively closed eyelids, and other factors, it is possible to receive insufficient ocular information when scanning an iris image. This is, however, a reasonably reliable way of identification.

The so-called Hamming distance, or the amount of bits in which the contrast of two diverse iris designs differs, represents the precision of comparing the two iris patterns. The Hamming distance is 0.32 for a chance of an inaccurate assessment of 1:26,000,000, according to reports.

When associating a large number of irises, Figure 4.21 displays the supply of Hamming's distance (Lizoul et al., 2012). A binomial distribution with a chance of 0.5 is seen in the graph. The graph also shows that two separate irises differing in < one-third of the info are exceedingly implausible.

4.4.2.2. Retina

The accuracy of retinal scanning is very great. Nevertheless, there are several circumstances in which obtaining a good image of the retina is impossible. It is depraved illumination in particular—because of the enormous amount of light, the user has a heavily closed pupil when scanning. Another issue arises as a result of the disorders above or other ocular dysfunctions.

The use of the retina for recognition is not particularly common, perhaps because there isn't much objective testing of this method. Sandia National Laboratory, an international corporation, verified EyeDentify Inc. on numerous hundred participants in 1991. The outcome was an untrue accept rate of nil and a false reject rate of < 1% (Potamianos & Maragos, 1996). Nevertheless, because biometric system testing was still in its early phases at the time, we cannot be certain of the test's neutrality.

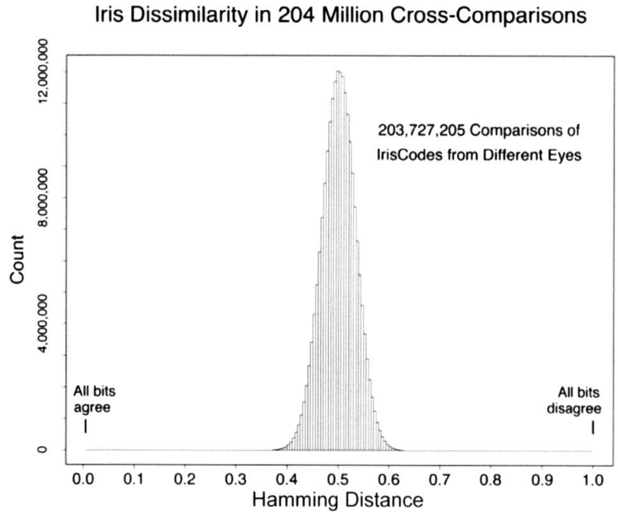

Figure 4.21. Hamming distance distribution.

Source: *https://www.cl.cam.ac.uk/~jgd1000/UAEdeployment.pdf.*

As stated by EyeDentify, the frequency circulation of every eye's image relative to the others advanced a near-perfect Gaussian curve along with a standard deviation of 0.176 and a mean of 0.144. With a specific mean value and a standard deviation of 0.7, the consistent possibility of this distribution is around one million (Elliott et al., 2004).

The wrong distance between the eye and the sensor, contact lens edges, unclean optics, and spectacles are all factors that might increase the false reject rate. Furthermore, because ambient brightness causes a subconscious narrowing of the pupil, the gadget cannot always be used outdoors during daylight hours.

4.4.3. Anti-Spoofing

4.4.3.1. Iris

There are numerous methods for determining whether the iris is alive. The most prevalent is the iris reaction to an alteration in light, in which the pupil shrinks with more intelligent lighting. This reflex is cataleptic, and replies normally take between 250 and 400 milliseconds. The pupil stretches and widens even under steady illumination, a phenomenon known as the hippus (Grother et al., 2009).

Blinking or eye movement on a scanning device's command can also be used to prevent spoofing.

More advanced devices employ the spectrographic characteristics of blood, lipids, and tissues. The iris pigment melanin, like blood, reflects infrared radiation very efficiently.

When light is reflected through a pink retina back into the camera, it is known as coaxial back retina reflection, often known as "red eyes."

Purkyne's replication of the cornea's surface and the lens can also be utilized to determine whether or not the eye is alive. Reflective pictures are formed when a sufficient light source illuminates the eye's surface, reflected from the front and rear surfaces of the lens and cornea.

4.4.3.2. Retina

Retinal scanning is a difficult and time-consuming operation that cannot be replicated. It would be essential to employ a deceived eye with the same properties as a real eye to fool such a sensor, which is difficult and near-impossible. There is not much info regarding the animateness test on the retina, but it might use medical data, such as the fact that the non-living retina is a diverse hue. It is also possible to examine the retina's refraction of light or bloodstream in blood vessels (Medina et al., 2011).

Because the eye is such a delicate organ, an invasive procedure cannot be employed. An alike aliveness test exists for the iris; nevertheless, when a phony eye replaces the right eye after a successful test life, this testing can be utilized to fraud the system. As a result, it is preferable to use a different method to check for liveness. The hue of the yellow spot is the first test. It is carried out while using the scanned eye. Only when a person dies does the yellow spot turn yellow; before then, it is reddish (Velisavljevic, 2009).

4.4.4. Related Standards

4.4.4.1. Iris

a) *ANSI INCITS 379–2004:* Iris Image exchange format defines the format for transferring iris image data. The description of attributes, sample tracking and data, and compliance criteria are all part of this process (Garea et al., 2018).

b) ISO/IEC 19794–6: 2011: Iris Image Data (Saraereh et al., 2020). Information Technology—Biometric Data Interchange

Formats—Part 6: Iris Image Data Provides two different data representation formats. The first is depend on uncompressed direct storage, whereas the second involves some preprocessing; nonetheless, the data is compacted and contains iris information.

4.4.4.2. Retina

In order to recognize the retina, there are no biometric values; nevertheless, they are essentially photographs of the bloodstream and hand vein detection. Thus similar criteria might be used. Only medical standards, such as ISO 10943:2011.

4.4.5. Commercial Applications and Devices

4.4.5.1. Iris

There are a plethora of practical applications to choose from. The most prevalent systems are found in seaports and airports around the United Arab Emirates. Another example is the method utilized by passengers with frequent flights in the Netherlands at Schiphol Airport. Another example is a Tokyo application. Employees of the condominium utilize this method to enter while a lift is summoned to transport them to their offices. The United Nations High Commission (UNHC) in Afghanistan utilizes iris recognition to keep track of immigration from neighboring nations (Richardson & Spivey, 2004).

4.4.5.2. Retina

In sectors where great security is required, like military and government locations, armaments development, industry, nuclear development, secret organizations, and so on, retinal recognition is acceptable.

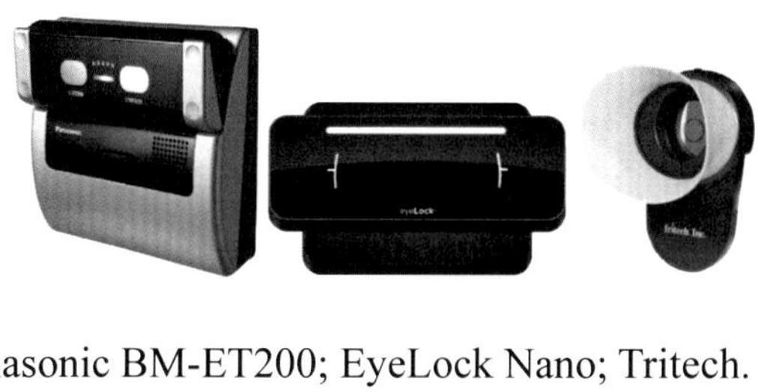

Figure 4.22. Panasonic BM-ET200; EyeLock Nano; Tritech.

Source: *https://www.researchgate.net/figure/Panasonic-BM-ET200-EyeLock-Nano-Iritech_fig1_327297061.*

To conclude this chapter, we focus on creating an intriguing and hitherto unknown technology that can be employed in both ophthalmology and biometric systems. This is a non-mydriatic fundus camera that is entirely automated. We started with a modest gadget many years ago, but as time went on, we progressed to the 3rd generation of the device. We are now developing the fourth version of this technology, which will be completely automated. The novel concept was solely engrossed on the retina, but we later arrived to repossess both the retina and the eye's iris in a device, whereas the 3rd and 4th generations are once again solely engrossed on the retina of the eye (Dimitriadis & Maragos, 2006).

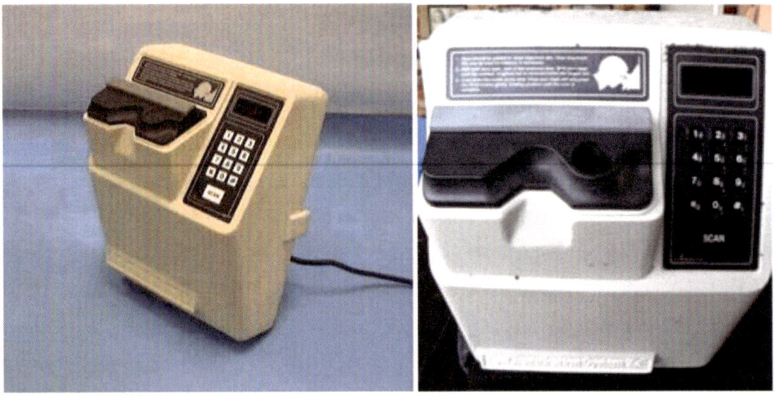

Figure 4.23. Identify 7.5 eye dentification system.

Source: *https://www.recycledgoods.com/eyedentify-7–5-eyedentification-system-biometric-retinal-identification-scanner/.*

Figure 4.24. A non-mydriatic fundus camera of our development—first generation on the left, the second generation in the middle, and third-generation on the right.

Source: *https://link.springer.com/chapter/10.1007/978-3-030–27731–4_11.*

The 3rd generation can now find the eye in the camera, shift the optical system to the image's, and invisible spectrum capture photos of the eye retina to record a brief video. The fourth-generation will record practically the full ocular background and integrate it into a single file (Wildes, 19197).

It will almost certainly be linked to software that can now locate the blind spot and macula, as well as vessels and arteries, extract and detect crossings and bifurcations, and locate areas with possible pathological findings, as well as to detect exudates/druses and hemorrhages, as well as calculate their area. In the future, we will concentrate on the precision and consistency of extractors and detectors and other sorts of illnesses that will be of primary concern to ophthalmologists (Bhagwagar & Rathod, 2015).

REFERENCES

1. Abdolahi, M., Mohamadi, M., & Jafari, M. (2013). Multimodal biometric system fusion using fingerprint and iris with fuzzy logic. *International Journal of Soft Computing and Engineering*, *2*(6), 504–510.
2. Abdullah, M. A., Dlay, S. S., Woo, W. L., & Chambers, J. A. (2016). A framework for iris biometrics protection: a marriage between watermarking and visual cryptography. *IEEE Access*, *4*, 10180–10193.
3. Abhyankar, A., & Schuckers, S. (2009). Iris quality assessment and bi-orthogonal wavelet based encoding for recognition. *Pattern Recognition*, *42*(9), 1878–1894.
4. Adeoye, O. S. (2010). A survey of emerging biometric technologies. *International Journal of Computer Applications*, *9*(10), 1–5.
5. Alsaadi, I. M. (2015). Physiological biometric authentication systems, advantages, disadvantages and future development: A review. *International Journal of Scientific & Technology Research*, *4*(12), 285–289.
6. Attebo, K., Mitchell, P., Cumming, R., & BMath, W. S. (1997). Knowledge and beliefs about common eye diseases. *Australian and New Zealand Journal of Ophthalmology*, *25*(3), 283–287.
7. Azimi, M., Rasoulinejad, S. A., & Pacut, A. (2019). Iris recognition under the influence of diabetes. *Biomedical Engineering/Biomedizinische Technik*, *64*(6), 683–689.
8. Beach, P., & McConnel, J. (2019). Eye tracking methodology for studying teacher learning: A review of the research. *International Journal of Research & Method in Education*, *42*(5), 485–501.
9. Bhagwagar, N. M., & Rathod, Y. A. (2015). A Survey on iris recognition for authentication. *International Journal of Technical Research and Applications*, *3*(2), 148–151.
10. Bhatia, R. (2013). Biometrics and face recognition techniques. *International Journal of Advanced Research in Computer Science and Software Engineering*, *3*(5), 4–19.
11. Bojko, A. (2006). Using eye tracking to compare web page designs: A case study. *Journal of Usability Studies*, *1*(3), 112–120.
12. Bouma, H., & Baghuis, L. C. J. (1971). Hippus of the pupil: Periods of slow oscillations of unknown origin. *Vision Research*, *11*(11), 1345–1351.

13. Brown, P. K., & Wald, G. (1964). Visual pigments in single rods and cones of the human retina. *Science*, *144*(3614), 45–52.
14. Bulling, A., Ward, J. A., Gellersen, H., & Tröster, G. (2010). Eye movement analysis for activity recognition using electrooculography. *IEEE Transactions on Pattern Analysis and Machine Intelligence*, *33*(4), 741–753.
15. Cheung, N., & Wong, T. Y. (2007). Obesity and eye diseases. *Survey of Ophthalmology*, *52*(2), 180–195.
16. Clark, M., Bovik, A. C., & Geisler, W. S. (1987). Texture segmentation using Gabor modulation/demodulation. *Pattern Recognition Letters*, *6*(4), 261–267.
17. Dabas, P., & Khanna, K. (2013). A study on spatial and transform domain watermarking techniques. *International Journal of Computer Applications*, *71*(14), 5–10.
18. Daugman, J. (2009). How iris recognition works. In *The Essential Guide To Image Processing, 12*(1), 21–30.
19. Daugman, J. G. (1993). High confidence visual recognition of persons by a test of statistical independence. *IEEE Transactions on Pattern Analysis and Machine Intelligence*, *15*(11), 1148–1161.
20. de Luis-García, R., Alberola-Lopez, C., Aghzout, O., & Ruiz-Alzola, J. (2003). Biometric identification systems. *Signal Processing*, *83*(12), 2539–2557.
21. Dimitriadis, D., & Maragos, P. (2006). Continuous energy demodulation methods and application to speech analysis. *Speech Communication*, *48*(7), 819–837.
22. Dobeš, M., Machala, L., Tichavský, P., & Pospíšil, J. (2004). Human eye iris recognition using the mutual information. *Optik*, *115*(9), 399–404.
23. Doherty, S., O'Brien, S., & Carl, M. (2010). Eye tracking as an MT evaluation technique. *Machine translation*, *24*(1), 1–13.
24. Drahanský, M., & Yang, J. (2018). Recognition of Eye Characteristics. In *Machine Learning and Biometrics*, 3(1), 7–35.
25. Durlu, Y. K. (2021). Response to treatment with intravitreal anti-vascular endothelial growth factors in bilateral exudative cuticular drusen. *American Journal of Ophthalmology Case Reports*, *22*, 101110.
26. Elliott, S. J., Peters, J. L., & Rishel, T. J. (2004). An Introduction to Biometrics Technology: Its Place in Technology Education. *Journal of Industrial Teacher Education*, *41*(4), n4.

27. Engerman, R. L. (1989). Pathogenesis of diabetic retinopathy. *Diabetes*, *38*(10), 1203–1206.
28. Fadool, J. M., & Linser, P. J. (1993). 5A11 antigen is a cell recognition molecule which is involved in neuronal-glial interactions in avian neural retina. *Developmental Dynamics*, *196*(4), 252–262.
29. Foster, N. E., Thomas, E., Hill, J. C., & Hay, E. M. (2010). The relationship between patient and practitioner expectations and preferences and clinical outcomes in a trial of exercise and acupuncture for knee osteoarthritis. *European Journal of Pain*, *14*(4), 402–409.
30. Garea-Llano, E., Osorio-Roig, D., & Hernandez-Hernandez, O. (2018). Image Quality Evaluation for Video Iris Recognition in the Visible Spectrum. *Biosens Bioelectron Open Acc: BBOA-144. DOI*, *10*, 2577–2260.
31. Grother, P., Tabassi, E., Quinn, G. W., & Salamon, W. (2009). Performance of iris recognition algorithms on standard images. *NIST Interagency Report*, *7629*, 1–120.
32. Hangai, M., He, S., Hoffmann, S., Lim, J. I., Ryan, S. J., & Hinton, D. R. (2006). Sequential induction of angiogenic growth factors by TNF-α in choroidal endothelial cells. *Journal of Neuroimmunology*, *171*(1–2), 45–56.
33. Hill, R. B. (1996). Retina identification. *Biometrics: Personal Identification in Networked Society*, 2, 123–141.
34. Holsanova, J. (2014). Reception of multimodality: Applying eye tracking methodology in multimodal research. *Routledge Handbook of Multimodal Analysis*, 2, 285–296.
35. Hornof, A., Cavender, A., & Hoselton, R. (2003). Eyedraw: a system for drawing pictures with eye movements. *ACM SIGACCESS Accessibility and Computing*, (77–78), 86–93.
36. Jacob, R. J. (1991). The use of eye movements in human-computer interaction techniques: what you look at is what you get. *ACM Transactions on Information Systems (TOIS)*, *9*(2), 152–169.
37. Jacob, R. J. (1993). Eye movement-based human-computer interaction techniques: Toward non-command interfaces. *Advances in Human-Computer Interaction*, *4*, 151–190.
38. Jain, A. K., Ross, A., & Prabhakar, S. (2004). An introduction to biometric recognition. *IEEE Transactions on Circuits and Systems for Video Technology*, *14*(1), 4–20.

39. Jain, A., Hong, L., & Pankanti, S. (2000). Biometric identification. *Communications of the ACM*, *43*(2), 90–98.

40. Jauregui, R., Cho, G. Y., Takahashi, V. K., Takiuti, J. T., Bassuk, A. G., Mahajan, V. B., & Tsang, S. H. (2018). Caring for hereditary childhood retinal blindness. *The Asia-Pacific Journal of Ophthalmology*, *7*(3), 183–191.

41. Jun, B. H., Noh, M. S., Kim, J., Kim, G., Kang, H., Kim, M. S., ... & Lee, Y. S. (2010). Multifunctional silver-embedded magnetic nanoparticles as SERS nanoprobes and their applications. *Small*, *6*(1), 119–125.

42. Kempen, J. H., O'Colmain, B. J., Leske, M. C., Haffner, S. M., Klein, R., Moss, S. E., ... & Hamman, R. F. (2004). The prevalence of diabetic retinopathy among adults in the United States. *Archives of Ophthalmology* (Chicago, Ill.: 1960), *122*(4), 552–563.

43. Krapichler, C., Haubner, M., Engelbrecht, R., & Englmeier, K. H. (1998). VR interaction techniques for medical imaging applications. *Computer Methods and Programs in Biomedicine*, *56*(1), 65–74.

44. Kronfeld, P. C. (1962). The gross anatomy and embryology of the eye. In *Vegetative Physiology and Biochemistry*,12, 1–62.

45. Lee, P. P., Feldman, Z. W., Ostermann, J., Brown, D. S., & Sloan, F. A. (2003). Longitudinal prevalence of major eye diseases. *Archives of Ophthalmology*, *121*(9), 1303–1310.

46. Lim, S., Lee, K., Byeon, O., & Kim, T. (2001). Efficient iris recognition through improvement of feature vector and classifier. *ETRI Journal*, *23*(2), 61–70.

47. Lizoul, K., André, H., & Guillet, F. (2012). Spectral precision of frequency demodulation method: Influence of additive noise on instantaneous angular speed spectral estimation. *Mechanical Systems and Signal Processing*, *164*, 108178.

48. Lu, Y., He, X., Wen, Y., & Wang, P. S. (2014). A new cow identification system based on iris analysis and recognition. *International Journal of Biometrics*, *6*(1), 18–32.

49. Lykins, A. D., Meana, M., & Kambe, G. (2006). Detection of differential viewing patterns to erotic and non-erotic stimuli using eye-tracking methodology. *Archives of Sexual Behavior*, *35*(5), 569–575.

50. Ma, L., Tan, T., Wang, Y., & Zhang, D. (2004). Efficient iris recognition by characterizing key local variations. *IEEE Transactions on Image Processing*, *13*(6), 739–750.

51. Majumder, S., Devi, K. J., & Sarkar, S. K. (2013). Singular value decomposition and wavelet-based iris biometric watermarking. *IET Biometrics*, *2*(1), 21–27.
52. Mallat, S. G. (1989). Multifrequency channel decompositions of images and wavelet models. *IEEE Transactions on Acoustics, Speech, and Signal Processing*, *37*(12), 2091–2110.
53. Martin, K., Lu, H., Bui, F. M., Plataniotis, K. N., & Hatzinakos, D. (2009). A biometric encryption system for the self-exclusion scenario of face recognition. *IEEE Systems Journal*, *3*(4), 440–450.
54. Medina, J. M., Pereira, L. M., Correia, H. T., & Nascimento, S. M. (2011). Hyperspectral optical imaging of human iris in vivo: characteristics of reflectance spectra. *Journal of Biomedical Optics*, *16*(7), 076001.
55. Mele, M. L., & Federici, S. (2012). Gaze and eye-tracking solutions for psychological research. *Cognitive Processing*, *13*(1), 261–265.
56. Metzner, H. J., Weimer, T., Kronthaler, U., Lang, W., & Schulte, S. (2009). Genetic fusion to albumin improves the pharmacokinetic properties of factor IX. *Thrombosis and Haemostasis*, *102*(10), 634–644.
57. Mohamed, Q., Gillies, M. C., & Wong, T. Y. (2007). Management of diabetic retinopathy: a systematic review. *Jama*, *298*(8), 902–916.
58. Monro, D. M., Rakshit, S., & Zhang, D. (2007). DCT-based iris recognition. *IEEE Transactions on Pattern Analysis and Machine Intelligence*, *29*(4), 586–595.
59. Mothi, R., & Karthikeyan, M. (2019). Protection of bio medical iris image using watermarking and cryptography with WPT. *Measurement*, *136*, 67–73.
60. Murawski, K., & Różanowski, K. (2013). Pattern Recognition Algorithm for Eye Tracker Sensor Video Data Analysis. *Acta Physica Polonica, A.*, *124*(3), 3–10.
61. Norn, M. (1985). Meibomian orifices and Marx's line studied by triple vital staining. *Acta Ophthalmologica*, *63*(6), 698–700.
62. Pan, L., & Xie, M. (2005, October). Research on iris image preprocessing algorithm. In *IEEE International Symposium on Communications and Information Technology,* 1, 161–164.
63. Paunwala, M., & Patnaik, S. (2014). Biometric template protection with DCT-based watermarking. *Machine Vision and Applications*, *25*(1), 263–275.

64. Potamianos, A., & Maragos, P. (1994). A comparison of the energy operator and the Hilbert transform approach to signal and speech demodulation. *Signal Processing*, *37*(1), 95–120.
65. Potamianos, A., & Maragos, P. (1996). Speech formant frequency and bandwidth tracking using multiband energy demodulation. *The Journal of the Acoustical Society of America*, *99*(6), 3795–3806.
66. Pueyo, V., Ara, J. R., Almarcegui, C., Martin, J., Güerri, N., García, E., ... & Fernandez, F. J. (2010). Sub-clinical atrophy of the retinal nerve fibre layer in multiple sclerosis. *Acta Ophthalmologica*, *88*(7), 748–752.
67. Raja, K. B., Raghavendra, R., Vemuri, V. K., & Busch, C. (2015). Smartphone based visible iris recognition using deep sparse filtering. *Pattern Recognition Letters*, *57*, 33–42.
68. Richardson, D. C., & Spivey, M. J. (2004). Eye tracking: Characteristics and methods. *Encyclopedia of Biomaterials and Biomedical Engineering*, *3*, 1028–1042.
69. Roberts, J. E. (2005). Update on the positive effects of light in humans. *Photochemistry and photobiology*, *81*(3), 490–492.
70. Ross, A., & Othman, A. (2010). Visual cryptography for biometric privacy. *IEEE Transactions on Information Forensics and Security*, *6*(1), 70–81.
71. Sabhanayagam, T., Venkatesan, V. P., & Senthamaraikannan, K. (2018). A comprehensive survey on various biometric systems. *International Journal of Applied Engineering Research*, *13*(5), 2276–2297.
72. Sadikoglu, F., & Uzelaltinbulat, S. (2016). Biometric retina identification based on neural network. *Procedia Computer Science*, *102*, 26–33.
73. Saraereh, O. A., Alsaraira, A., Khan, I., & Choi, B. J. (2020). A hybrid energy harvesting design for on-body internet-of-things (IoT) networks. *Sensors*, *20*(2), 407.
74. Sarode, N. S., & Patil, A. M. (2014). Review of iris recognition: an evolving biometrics identification technology. *International Journal of Innovative Science and Modern Engineering (IJISME)*, *2*(10), 34–40.
75. Scanlon, P. H., Wilkinson, C. P., Aldington, S. J., & Matthews, D. R. (2009). Classification of diabetes. *A Practical Manual of Diabetic Retinopathy Management*, *17*, 2–10.

76. Shaw, A. K., Majumder, S., Sarkar, S., & Sarkar, S. K. (2013). A novel EMD based watermarking of fingerprint biometric using GEP. *Procedia Technology*, *10*, 172–183.
77. Simon, C. (1935). A new scientific method of identification. *New York state Journal of Medicine*, *35*(18), 901–906.
78. Soliman, R. F., El Banby, G. M., Algarni, A. D., Elsheikh, M., Soliman, N. F., Amin, M., & Abd El-Samie, F. E. (2018). Double random phase encoding for cancelable face and iris recognition. *Applied Optics*, *57*(35), 10305–10316.
79. Stanetić, K. D., Savić, S. M., & Račić, M. (2016). The prevalence of stress and burnout syndrome in hospital doctors and family physicians. *Medicinski pregled*, *69*(11–12), 356–365.
80. Sukumaran, S., & Punithavalli, M. (2009). Retina recognition based on fractal dimension. *IJCSNS Int J Comput Sci and Netw Secur*, *9*(10), 66–7.
81. Tan, C. W., & Kumar, A. (2014). Accurate iris recognition at a distance using stabilized iris encoding and Zernike moments phase features. *IEEE Transactions on Image Processing*, *23*(9), 3962–3974.
82. Tan, C. W., & Kumar, A. (2014). Efficient and accurate at-a-distance iris recognition using geometric key-based iris encoding. *IEEE Transactions on Information Forensics and Security*, *9*(9), 1518–1526.
83. Timberlake, G. T., & Kennedy, M. (2005). The direct ophthalmoscope how it works and how to use it. *University of Kansas*, *39, 4–10.*
84. Tower, P. (1955). The fundus oculi in monozygotic twins: report of six pairs of identical twins. *AMA Archives of Ophthalmology*, *54*(2), 225–239.
85. Trokielewicz, M., Czajka, A., & Maciejewicz, P. (2014, November). Cataract influence on iris recognition performance. In *Photonics Applications in Astronomy, Communications, Industry, and High-Energy Physics Experiments, 9290*, 929020.
86. Trokielewicz, M., Czajka, A., & Maciejewicz, P. (2020). Post-mortem Iris Decomposition and its Dynamics in Morgue Conditions. *Journal of Forensic Sciences*, *65*(5), 1530–1538.
87. Turk, M., & Pentland, A. (1991). Eigenfaces for recognition. *Journal of Cognitive Neuroscience*, *3*(1), 71–86.
88. Vander Heiden, M. G., Li, X. X., Gottleib, E., Hill, R. B., Thompson, C. B., & Colombini, M. (2001). Bcl-xL promotes the open configuration

of the voltage-dependent anion channel and metabolite passage through the outer mitochondrial membrane. *Journal of Biological Chemistry, 276*(22), 19414–19419.

89. Vatsa, M., Singh, R., Mitra, P., & Noore, A. (2004, October). Digital watermarking based secure multimodal biometric system. In *2004 IEEE International Conference on Systems, Man and Cybernetics, 3,* 2983–2987.
90. Velisavljevic, V. (2009). Low-complexity iris coding and recognition based on direction lets. *IEEE Transactions on Information Forensics and Security*, *4*(3), 410–417.
91. Venkitaraman, A., & Seelamantula, C. S. (2014). Binaural signal processing motivated generalized analytic signal construction and AM-FM demodulation. *IEEE/ACM Transactions on Audio, Speech, and Language Processing*, *22*(6), 1023–1036.
92. Von Stetina, S. E., Watson, J. D., Fox, R. M., Olszewski, K. L., Spencer, W. C., Roy, P. J., & Miller, D. M. (2007). Cell-specific microarray profiling experiments reveal a comprehensive picture of gene expression in the C. elegans nervous system. *Genome Biology*, *8*(7), 1–32.
93. Wildes, R. P. (1997). Iris recognition: an emerging biometric technology. *Proceedings of the IEEE*, *85*(9), 1348–1363.
94. Wilkinson, C. P., Ferris III, F. L., Klein, R. E., Lee, P. P., Agardh, C. D., Davis, M., ... & Group, G. D. R. P. (2003). Proposed international clinical diabetic retinopathy and diabetic macular edema disease severity scales. *Ophthalmology*, *110*(9), 1677–1682.
95. Yau, J. W., Rogers, S. L., Kawasaki, R., Lamoureux, E. L., Kowalski, J. W., Bek, T., ... & Meta-Analysis for Eye Disease (META-EYE) Study Group. (2012). Global prevalence and major risk factors of diabetic retinopathy. *Diabetes Care*, *35*(3), 556–564.
96. Zhang, X., Saaddine, J. B., Chou, C. F., Cotch, M. F., Cheng, Y. J., Geiss, L. S., ... & Klein, R. (2010). Prevalence of diabetic retinopathy in the United States, 2005–2008. *Jama*, *304*(6), 649–656.

Chapter 5

Machine Learning in Biometric Signature Verification

CONTENTS

5.1. INTRODUCTION

In electronic systems and pattern recognition, biometric systems have seen a tremendous growth in terms of their adoption. Biometric systems are security systems that have the potential to recognize or confirm the individual according to their behavioral physiological characters. The word *biometric* is the combination of two Greek words: bio and metric. The meaning of *Bio* is life and the *metric* is to estimate or measure. In the 1970s, to recognize criminals with the help of fingerprints, biometric systems were mainly offered for law enforcement agencies (Chadha et al., 2013).

Even though for authentication and recognition, biometric systems work with exclusive traits of a person, therefore it can be divided into two large features i.e., behavioral and physiological features (Vargas et al., 2007). Behavioral characteristics are usually variable that may vary with the course of time, age, mood, and other features. The means of behavioral characteristics include voice, keystroke, gait, signature, handwriting. On the contrary, physiological characteristics include special features of individuals and do not change with time, such as thumb impression, fingerprints, iris recognition, palm print, DNA, veins, and face recognition (Rosso et al., 2016). Behavioral and physiological traits are described in Figure 5.1.

An identification system or a verification system is the base for any biometric information system. The confirmation of the uniqueness of a person based on one's professed distinctiveness is the objective of this verification system. On the other hand, the identification system aims to generate the individual's distinctiveness (among humans registered in the expert system) without the person showing their description. Put simply, an identification system is a one-to-many (1:N) search system while a verification system is a one-to-one (1:1) search system (Bashir et al., 2015). To improve the security levels and act as a satisfactory vital part in a group of methods like writer identification, bank check processing, face detection, online banking, medical detection, security purposes, attendance, and official document, etc., these biometric systems are performing a key role in diverse areas (Rosso et al., 2016).

Although there are various biometric systems, signature verification is one of the most leveraging and demanding behavioral biometric systems. The term signature is extracted from the source of Latin *signer* means to sign. In someone's writing, any handwritten specimen that describes the person's surname, first name, nickname, last name, or grouping of them used for recognition is the signature (Durrani et al., 2016). The procedure of discriminating forged and genuine signers from the pre-stock recognized source samples is the system of signature verification. Along with huge progress in several applications, due to the leading properties like non-invasive, user-friendliness, socially and legally unobjectionable by the society, it is one of the most suitable biometric verification systems.

In this modern age, the verification of handwritten signatures has been broadly considered. A technique of signature verification can be characterized into two key modules: offline and online signature verification. By taking the image of the signature using smartphones, tablets, magnetic pads, PDA's, the sequential data is measured by the online signature verification system. It is usually variable that works on dynamic characteristics like positions of pen tips, the direction of writing, order of strokes, velocity, pressure, and speed, etc. (Alizadeh et al., 2010).

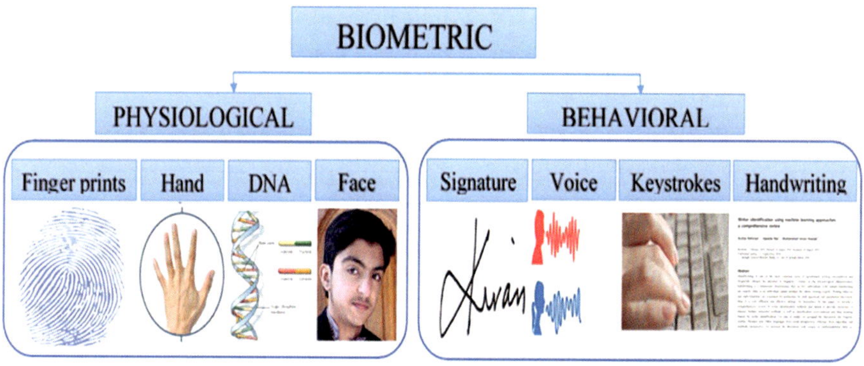

Figure 5.1. Behavioral and physiological features of biometric.

Source: https://in.pinterest.com/pin/558164947544149473/.

Instead, the verification system of an offline signature utilizes an optical detector to get the signature on the page, document, or image is scanned. This technique covers static information like height and length of the signature, slant, baseline, bounding box, pressure, and size as it is static (Patil & Hegadi, 2013). Due to the lacking of dynamic constraints detail, verification of the offline signature is comparatively harder than verification of the online signature. So higher identification rates and better precision are provided by verification systems of online signature than verification systems of offline signature. Figure 5.2 shows the working plan of verification systems of online and offline signature.

Distinguish the signature modules based on differences, is the main aim of the verification system of signature (Fayyaz et al., 2015). There are like Forged and genuine signatures are eventually two different groups of signatures. *A* signature that owns by an original individual is genuin*e*. *Forged signature* comprises the imitated or copied signature of an individual formed by an illegal person. The identification of the forged signature to decrease hacking information and crimes is one of the main features of the signature verification system. So the kind of signature forgery is the result of the methods of the signature verification system (Kalenova, 2003). Forgeries of the signature can be more characterized into three groups like skilled forgery random forgery, simple forgery.

In *zero effort* or random forgery, an individual does not have data about the signature shape or name of the original individual. *Casual forgery/simple forgery* is an amateur forgery in which the knowledge of signature shape is unidentified but the name of the original signatory identifies. . Though, qualified copying of the genuine signature is a *simulated or skilled forgery*. In this kind of forgery, information about the name of the original signer and its shape is known therefore it is harder to detect (Hanmandlu et al., 2005). Other than these forgery classes, other classes had been presented like freehand signature forgery, unskilled forgery, electronic forgery, tracing forgery, targeted forgery, cut-and-paste forgery. There is less focus on these forgeries but data related to them can be determined from the work of Nguyen et al. (2007) and Deore et al. (2015)

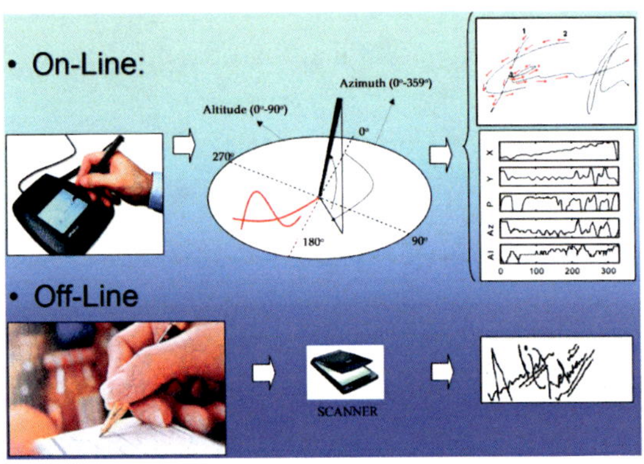

Figure 5.2. The sequence of the online and off-line verification of the signature.

Source: *https://www.researchgate.net/figure/Workflow-of-offline-and-online-signature–verification_fig2_335191144.*

This chapter aims to give a deep analysis of signature authentication systems to the readers. Unlike various feedbacks on this subject, this chapter gives extra information which is missing in other feedback articles, like:

a) Comprehensively, literature reviews and tabulate datasets statistics.
b) Give deep knowledge in the field of signature for the researchers about the datasets.
c) Critical examination assessment on the literature review of signature verification.
d) Covers modern market tasks along with future predictions for the verification system of signature.
e) Gives comprehensive background about verification of the signature.
f) According to the classification models' taxonomy, the evaluation of the verification system of signature.

5.2. BACKGROUND

The study of signature verification is an old idea that goes back to 439 AD when signatures were used for verification of documents in the Roman Empire. The Roman Empire was one of the earliest governments that had these kinds of legislations. In 1792 these frameworks started to appear in English-speaking territories. Until Common Law Procedure Act was accepted in England in 1854, these laws keep under progress. In previous old ages, the handwriting was physically confirmed and the technique itself is lucrative and boring. Far along in the 20th century, as the computer was designed, to trace the limitations of conservative methods, the analyst tried to arrange applications to confirm the handwritten signature. To explain or investigate the signatures and handwriting, automated signature recognition is the method to deal with steady machines.

Since 1960, in information technology, signature verification is emerging quickly (Fayyaz et al., 2015). In 1965, the first automated recognition system of signature was introduced in North America Aviation. The first online and offline signature verification system was introduced by Liu and Herbst (1977) and Nagel and Rosenfeld (1973). To handle this issue, various algorithms and methods have been suggested in the following of these studies. The first work in online signature identification was published by Mauceri in 1965. In 1966, for the offline verification system, the first time computer was used by Kozinets et al. (2016). In 1973, in IEEE conferences and transactions, the first verification algorithms of offline signature were published by Rosenfeld and Nagel (1973). Similarly, in 1977, the verification algorithm of online signature was published by Herbst and Liu (1977). In 1986, to take out information related to pressure from grayscale images, the process of Pseudo-dynamic feature extraction was proposed. The first study for the suggested technique was presented by Plamondon and Sabourin in 1987. In the recognized case of Daubert vs. Merrell Dow Pharmaceuticals, the approach was verified in 1993. In 2004, for the first time, a worldwide competition for verification of online signature was directed by the Hong Kong University.

5.3. STAGES OF THE SIGNATURE VERIFICATION SYSTEM

There are various conventional stages of the verification system of signature but at first, it is clear that method is offline verification of signature or online verification of the signature. The conventional stages of signature verification consist of feature extraction, pre-processing, data acquisition, and classification. Figure 5.3 realistically shows these core stages of the verification system of signature.

5.3.1. Data Acquisition

Taking the image of the signature is data acquisition (Deore & Handore, 2015). For any confirmation technique, data acquisition is the main step. The verification system of the handwritten signature can be characterized as either: online (dynamic) system or an offline (static) system because of data acquisition. Signatures were gathered with the help of gel and sketch pens and ball pens having blue and black ink on A4 paper and correspondingly each person gives different trials of his/her signatures for the offline signature. Then with an appropriate resolution, these signatures are then transformed into grayscale. After the completion of the writing process, data acquisition is executed by using scanners and cameras. Then in the database in JPEG or JPG or PNG format, these scanned pictures are initialized and kept. The forged and genuine signatures that use to test and train the model, are stored in every set of databases.

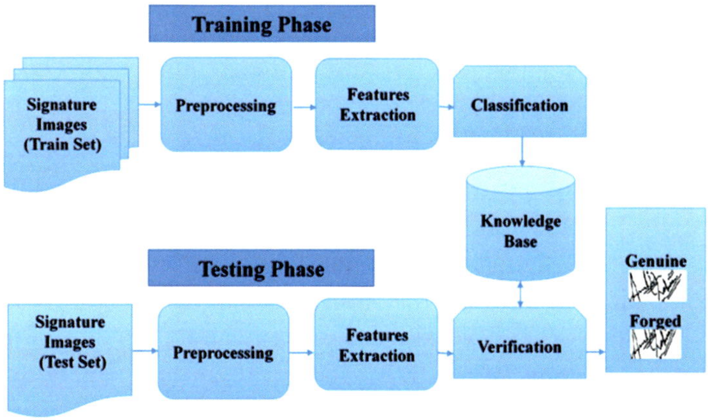

Figure 5.3. Roadmap of the verification system of signature.

Source: *https://www.researchgate.net/figure/Workflow-of-offline-and-online-signature–verification_fig2_335191144.*

On the other hand, in the verification system of online signature, for signature acquisition, graphical tablets, touch mobiles, and pen tablets are used. The digitizing tablets are the frequently used data acquiring devices. These specific gadgets generate a complete process of writing, electric signals, and features of the signature trace. Pressure, velocity, position, force signals, and acceleration are normally created signals. The signatures cannot be investigated for more processing once it generates. It requires a few pre-processing that will be described in the next segment (Durrani et al., 2016).

5.3.2. Pre-Processing

The simplification of some transactions without overlooking important processing is pre-processing (Bashir et al., 2015). To arrange samples of raw data in a typical form or to improve the input data that is suitable for the feature extraction phase, pre-processing is the basic step.

Pictures have low contrast, noisy pixels, blurriness, and complex backgrounds in the verification system of offline signature. So to develop an improved image that will be suitable for further activities, different pre-processing methods were organized. Background elimination, greyscale conversion, signature extraction, width normalization, noise removal by using filters, signature size normalization, signature alignment, binarization, skeletonization, thinning, and cropping, etc. are the common steps of pre-processing (Vargas et al., 2011).

While in the verification system of online signature, images may consist of various kinds of jerks and variations that require to be taken out in pre-processing stage. The process of preprocessing can be performed via smoothing algorithms, filtering, and noise reduction.

Pre-processing of verification of the offline signature is comparatively more complicated and time taking than the verification system of online signature. Signature segmentation might be the core reason because of the similarity of different letters, loops, differences in handwriting, overlapping of characters, etc. In offline signature verification systems. Put simply, the verification systems of online signature are more effective and precise because of pre-processing approaches (Van et al., 2007).

5.3.3. Features Extraction

The process of extracting the typical features of signatures that are used for the distinction of different groups of signatures is feature extraction (Al-Omar et al., 2011). The main factor of an effective system is the ideal selection of the leading set of features. From several different viewpoints, the verification system of the signature has been determined, developing many substitutes for the feature extraction. In general, the feature extraction approaches can be categorized as online/dynamic or offline/static. From the process of signature execution, the dynamic information is extracted in online/dynamic features like acceleration, position, velocity, and pressure, etc. While in offline/static features, the static information is extracted from the signature images such as signature slant, height, density, width, baseline, etc. There are three wide classes of features based on offline or online approaches: automatic features, statistical features, structural features, or model-based features (Rashidi et al., 2012). Statistical features are related to the statistical or mathematical measurements for the arrangement of appropriate information for decreasing the distance between different classes.

Additional wide classification of statistical features is between Local and Global Features. The representation of the whole image or the entire image of the signature is related to the global feature. Compactness, displacement, signature global orientation acceleration, area, total no. of components, time duration of positive or negative position, total signing duration, no. of pens ups/downs, coefficients obtained using mathematical transforms, etc. are included in common global features. In local feature extraction, from a particular part of an image, the features are extracted and the signature image is classified into several parts or segments. Pixel-oriented features and Component-oriented features are two factors of Local features (Iranmanesh et al., 2013). Component-oriented feature extracts the parameters at a component level such as geometric features, strokes position, height or width ratio of the stroke, the orientation of strokes, orientation-based features, slant, moment-based, and contour-based features, etc. Contrary to, At every level of the pixel, the pixel-oriented features are attained, for instance, texture features grid-based information, gray-level intensity-based features pixel consistency, gray level intensity, shadow code-based features and shape-based features, etc. (Ahmed et al., 2019).

Signatures' local structure includes structural features such as Scale-Invariant Feature Transform (SIFT), Local Binary Patterns (LBP), Speeded Up Robust Features (SURF) and pixel density within signature grids, etc. In previous years, approaches that do not depend on hand-crafted features have gained attention. Through particular models like Extend Learning Model (ELM), Convolutional Neural Network (CNN), Recurrent Neural Network (RNN), etc., the automatic or model-based features are learned from primary data (pixels, in the case of images (Ahmad et al., 2015).

5.3.4. Classification

The procedure to estimate the validity of the query signature is classification. The objective of classification is to corresponding the query signatures feature with the pre-stored information base features because of verification utilizing a huge quantity of reference signature-based examples in the train set. The query signature is characterized as forged or genuine after the training. Support Vector Machine (SVM), Rule-based approach, template matching, Neural Network (NN), etc. are the common claes in the domain of verification system of signature (Alizadeh et al., 2010).

The conventional phases of the verification system of a signature have been discussed, now distinguished previous research contributions to this area will be described in the coming segment.

5.4. A REVIEW OF SIGNATURE VERIFICATION SYSTEMS

In pattern recognition, the most dynamic issue is signature verification. As mentioned earlier, based on data acquisition the methods proposed for the signatures verification are usually categorized into two classes: online signature verification and offline signature verification. In the subgroups, by using different methods, a deep review of online signature verification and offline signature verification is explained (Ansari et al., 2014).

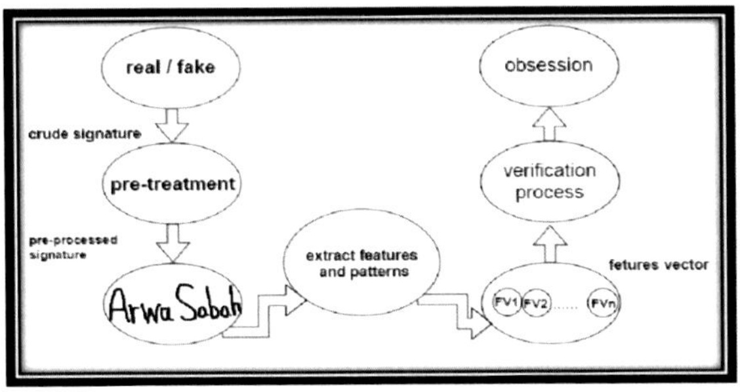

Figure 5.4. Flow diagram of verification system of a signature.

Source: *https://www.researchgate.net/figure/Flowchart-of-a-Signature–verification-System-Biometric-Measures-Biometric-systems-can_fig1_335568823.*

5.4.1. Offline Signature Verification

For attaining the best precision, high performance, and effectiveness in the domain of offline signature verification, numerous researchers applied several methods and procedures. Three common types of methods proposed for signature verification; statistical approaches, structural approaches, and template matching. We describe the different verification approaches developed under each one of the following groups (Bibi et al., 2020).

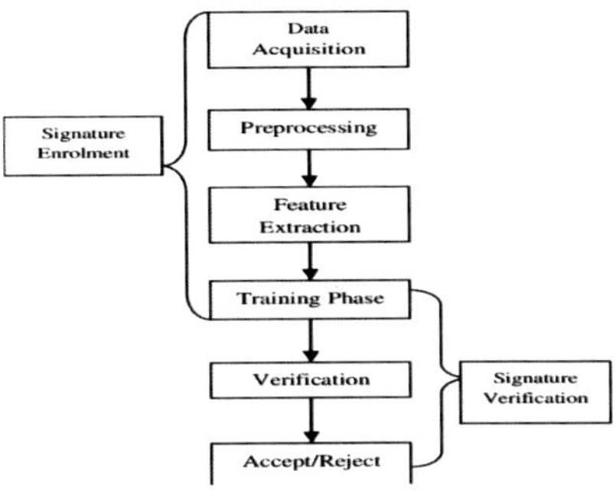

Figure 5.5. Block diagram of verification system of offline signature.

Source: *https://www.researchgate.net/figure/Block-Diagram-of-Offline-Signature–verification-System_fig1_272863797.*

5.4.1.1. Template Matching

The procedure of comparing template/figure in which test signatures are compared with pre-stored templates in the database. For this aim, the frequently used algorithm is Dynamic Time Warping (DTW). Here few of the short-term feedbacks of the researcher's contribution organizing DTW are mentioned.

Parziale et al. (2019) performed experiments on BiosecureID-SONOF and MCYT-100. They extracted eight statistical features were extracted by them such as pressure, acceleration, pen position, velocity, and pressure derivative. For comparing feature vectors, Euclidean distance was used and Z-score normalization was organized. Lastly, through DTW with two kinds of normalization, the arrangement was accomplished. By using the MCYT-100 database, an EER of 3.09% on skilled forgery and 1.30% on random forgery was attained. On the BiosecureID-SONOF dataset, the stated EER was 1.45% on skilled forgery and1.17% on random forgery.

By using the 2640 samples of signatures of the CEDAR Database, the signatures were confirmed by Kumar (2014). To extract the invariant features, the trace transform was organized by them. Through six circus functions, a feature vector of six features was developed and then standardization is accomplished. By utilizing a threshold value of 0.069, the similarity measures for different-writer and same-writer pairs were accomplished. The FAR, FRR, and ERR of 24.58%, 25.83%, and 24.4% correspondingly were stated by them.

Depending on DTW, a verification system of signature was showed by Bhunia et al. (2019). Firstly, vertical projection features were extracted and signatures samples were preprocessed. The DTW was improved and on the database of 100 individuals, classification was accomplished by them. An ERR of 2% and 29% is realized correspondingly.

For verification of a signature, an algorithm of a 2D geometric warp was applied by Kennard et al. (2012). Through an automatic morphing correspondence algorithm, a 2D geometric warp was figured to line up the strokes of an observed signature with reference. By utilizing the distance measures, the signatures were confirmed by them. The gained EER correctness on the Chinese dataset was 74% and the English dataset of blind forgeries had the EER of 94%-96%, 87%-91% on simple forgeries.

5.4.1.2. Statistical Approaches

The methods used for the statistical analysis of data gained from the statistical features of signature pictures are statistical approaches. To understand the difference and relation within the knowledge of two or more signatures, statistical approaches were used by few researchers. Hidden Markov Model (HMM), Genetic Algorithm (GA), distance measures, Random Forest, and Multilayer Perceptron (MLP) are a few of the statistical features in the area of verification of the signature.

5.4.1.3. SVM

Support Vector Machine (SVM) is a well-known statistical classifier where the main feature is a separating hyperplane. Put simply, the categorized training data is characterized with an ideal hyperplane algorithm. In two-dimensional space, this hyperplane is a line separating the plane into two parts while every category is put on any side. Similar to other statistical methods, for the researchers of the area, SVM was also the center of attention. In this segment, the most prominent work of SVM is explained. Through Taylor Series Expansion (TSE) the features were extracted by Shekar et al. (2019). They accomplished classification through SVM and used the MUKOS and CEDAR datasets. An accurateness of 98.93% on the MUKOS dataset and 95.25% on the CEDAR dataset correspondingly was regained by them.

An experiment on MCYT, CEDAR, and GPDS synthetic databases was performed by Sharif et al. (2020). In pre-processing stage, normalization, segmentation, morphological operations, binarization, and median filter methods were applied. Global, horizontal, and vertical features were extracted. With the help of Genetic algorithm, the best features were chosen. For the signatures verification, SVM was used. Through GPDS synthetic dataset FRR, FAR, and AER of 4.16%, 3.33%, 5.75%, by using CEDAR dataset 4.67%,4.67%,4.67%, and 3.67,6,67,5.0% through MCYT dataset respectively was stated by them. In preprocessing, contrast improvement and binarization were conducted by Okawa (2018). For feature extraction, VLAD encoding with KAZE features and the BoVW model was organized by them. With the stated error rate of 1.0% and 6.4%, an experiment was conducted on CEDAR and MCYT-75 datasets.

Through SVM and Radon transform, the verification of offline signature was explained by Kiani et al. (2009). To divide signature images, local windows were used. Usin Radon transform, line segment detection was measured to calculate the image intensity. By generating a feature vector for a line breadth of 6 pixels, extraction of a feature vector was designed. By segmenting the elements to the maximum value, normalization of the feature vector was carried out. For organizing the signature images, SVM was used. For verification that contains 30 signatures per 20 classes, a Persian signature dataset was used. 96% accuracy with FAR of 17% and FRR of 4% was attained. For training, the Stellenbosch dataset is used which consists of 22 classes, 10 genuine signatures, and for testing, 20 genuine signatures are used. FRR of 19% reached.

Otsu binarization was used to binarize the images (Kiani et al., 2009). With the use of radon transform that calculates the estimated sum of the image intensity concerning angle and line feature vector was created. With the comparison of peak value with a threshold, authentication of line segment existence was considered. Stellenbosch dataset that consists of 22 classes, 10 genuine signatures is also used by them. 19% FRR was reached. For the arrangement of offline signature, discrete Daubechies (db4) wavelet transform was suggested by Hegadi and Patil (2013). For three positions that were vertical, horizontal, and diagonal, detail coefficient and wavelet coefficient were extracted. Standard GPDS databases were utilized. For the set of the training, 4 forgery and 4 genuine signatures had been utilized, and all the residual signatures of persons were utilized for the set of testing. The pre-process of signature images was carried out and through wavelet transform of discrete Daubechies (db4), features were extracted. 80 features are represented by each signature. The Sequential Minimum Optimization (SMO) method in SVM classifiers had been utilized for fast optimization. Matlab software had been utilized for operation. FRR and FAR were 10%and 13% respectively in the linear kernel while in the nonlinear kernel, FRR and FAR was 12%and 15%.

On Bengali and English scripts, the work was carried out by Pal et al. (2011). A database of 846 (454 English+392 Bengali) signature samples was used for the set of testing and training, 1800 (1100 English+700 Bengali) signature samples were used. For extraction of features, background and Foreground information was used. To calculate 400-dimension foreground/background of 400-dimensional gradient and under-sampled bitmap features, it first standardized the input images by resizing the image into126×126 pixels and 200×800, and similarly, a boundary box was created for samples. The gradient feature's images were divided into 16 overlapping blocks as well as the normalized 400-dimensional image under-sampled bitmap features were divided into 100 blocks. Similarly, direction features of 400 modified chain-code were achieved the chain code frequencies were calculated. A factor was considered for gradient features in the gray-scale local-orientation histogram of an image. Implementation of Gaussian filter was carried out to get 5×5×16 = 400 gradient features, as well as the implementation of Robert's filter, was carried out to get the gradient image in which gradient tangent arc is quantified into 16 locations. To measure the distance, the Euclidian distance measure was considered. For the recognition of signature, Nearest Neighbor (NN) techniques were used as classifiers. The outcomes represented that 99.41% of best outcomes were achieved with the use of the SVM classifier and 400-dimension chain-code direction features, additional, good outcomes were determined by 200-dimension under-sampled bitmaps features (Pal et al., 2011).

On 500 signatures datasets having 200 forged signatures (20 signatures every 10 for one person) and 300 signatures genuine (30 signatures every signer had 10signatures), the work was done by Randhawa et al. (2012). By employing HP-scan jet 5400c at 300dpi, the signatures were scanned. By skeletonizing, binarizing, grayscale conversion, and normalizing the samples, pre-processing of the dataset was done. In the newly introduced system, for the extraction of zone features Image Processing Toolbox (IPT) of Matlab 6.0 was used and Hu Moments mined 28 features. For the aim of testing, 100 signatures were used and for training, 400 signatures were used. For teaching the SVM, a database of signatures was used. After that, regarding FAR, FIR, and FRR, the accuracy of signature verification of the model has been estimated. So, in this paper, SVM defined with 9% accurateness effectively proves the off-line signature.

On the openly offered Chinese dataset and GPDS-300 dataset, the work was done by Chen et al. (2011). In writer-independent and writer dependent mode, three different pseudo-dynamic features are based on gray level: histogram of oriented gradients (HOG), local binary pattern (LBP), and gray level co-occurrence matrix (GLCM) were considered in the suggested system. To conduct classification, classifiers named Support Vector Machine (SVMs) based on writer was considered. In SVM, every database is classified into three classes, each class having 100 persons. For every signer, from random forgeries and reference signature, SVM was accomplished with the attained feature vectors. The presentation was estimated with the use of skilled forgeries of the CSD dataset Chinese signature and dataset of GPDS signature. 9.94% and 5.66% equal value of error was the achieved outcomes with the use of a dataset of Chinese and GPDS signature. For verification of a signature, a method based on SVM was introduced by Shet & Kruthi (2014).

With the use of sketch pens, balls, gel in discrete models, from various individuals, a total of 336 signatures were acquired. By employing methods of image processing such as complementing, binarization, thinning, filtering, and edge detection, the preprocessing of scanned images was used. With the use of samples' pre-processing, statistical features such as calculation of a number of loops, centroid, horizontal and vertical profile, center of gravity, and normalized area were mined and kept in a database. After that for the arrangement of signatures, SVM was used and the total arrangement error value of 7.16% was attained.

For testing, 30 forged signatures and for training, 24 genuine signatures were taken from every writer, GPDS300 Signature CORPUS database was used by Parodi & Gómez (2014). From a grid of circular shapes, features of graphometry were mined. In the form of DFT, rotation was plotted to get robustness. To organize samples, a classifier based on SVM was considered. 7.82% FRR, 0.49% FAR, and 4.21% EER were found in the suggested method. With the use of three dissimilar databases (GPDS750, MCYT-75, and GPDS100), gray level information-based verification technique of signatures was proposed by Vargas et al. (2011). For the elimination of background, the process of Posterization was considered. Then a breakdown of a single-level 2-D wavelet was carried out with the use of Matlab function DTW2. For enhancement and calculation of the universal effect of ink-type, wavelet analysis was conducted. Through GLCM, features of Statistical texture were mined. As compared to other techniques, the suggested technique was accomplished with the use of SVM and 13.50% ERR was stated.

Features of slope angle and slope based on zone and geometric features based on concentric squares were mined by Randhawa et al. (2013). Extracted features are in the set of features. For pre-processing of signature, Image Processing Toolbox (IPT) was used in MATLAB 7.0. For the activity of verification, in order to teach the SVM model, the software of the LIBSVM 3.0 was used. With three dissimilar kinds of the kernel function. SVM was considered. With the comparison of polynomial and linear kernels, the optimum results were presented by the SVM model based on Radial Basis Function (RBF). 1.66% FRR and 1.25% FAR were obtained by the suggested system. With the use of a 2D Gaussian filter, a method of new feature extraction based on a grid was shown by Blumenstein & Nguyen (2011). The database of the GPDS-960 corpus was considered. In pre-processing, signature contours were figured out. By employing SVM, the procedure of learning and classification was conducted. For arbitrary forgeries, 0.02% FAR and 13.90% AER were achieved for the suggested system.

With the signature's polar feature signifier, the verification system of offline signature was presented by Pushpalatha et al. (2013)., To get a regression score, PLS Regression was implemented on registered class signatures. Ather that by using Hidden Markov Model, the Log-Likelihood of the class signatures was measured.

The verification of the classification was considered authentic if the Log-Likelihood distance deviation and regression score was less than 5%. For verification, the machine of Multiclass Support Vector was considered. With 8%.

FAR, the correctness of the suggested method was 98%. A skilled forgery with a precision of 71% and arbitrary forgery with a precision of 76% was pointed out by the proposed technique. For the extraction of features, the method of G-SURF and global filtering was joint by Pal et al. (2012). On the database of GPDS, the technique was developed. By the use of SVM, the procedure of classification and learning was conducted. For the suggested technique, 2.35% FRR, 3.55% FAR, and 2.95% AER was attained.

Brazilian PUC-PR and GPDS-960, the work was done by Hafemann et al. (2019). For feature enrollment and learning, the dataset was divided into a development set and exploitation set. The CNN is considered a feature extractor. With the algorithm of OTSU, images were resized with the use of bilinear interpolation.

By filling the images with white background, the image's normalization was carried out. For the arrangement of writer-dependent, SVMs and linear SVMs with the RBF kernel were implemented. With achievement from 14.64% of EER to 10.70%, on a dataset of GPDS-160, normalization got good classification outcomes.

On the dataset of Brazilian, 9.83% EER with five signatures, and in the dataset of GPDS, 15.05% EER with four signatures were reported by them. For investigation, datasets of the Brazilian PUC-PR, CEDAR, MCYT-75, and GPDS-160 were considered.

They extracted and based The features based on CNN and handcrafted were mined by them and for sorting, added to the linear SVM. Two countermeasures on the achieved value of the attacks were calculated and to distinguish the forgeries, the biometric attacks were estimated by them.

5.4.1.4. Structural Approaches

To show the signature patterns, the structure of symbolic data, such as trees, strings, and graphs are considered in the structural approaches. The comparison of symbolic representation with prototypes will occur in this kind of system. To determine whether a correspondence alignment exists among two images, structural approaches are used.

The work on the database of Hindi signatures was done by Pal et al. (2017). 30 skilled forgeries and eight genuine signatures were used for testing and 16 genuine signatures were used for training. For binarization, mean filtering and histogram thresholding were implemented. Overall, 24 features were mined. For verification, a method of symbolic presentation of the value of inter-based was used. The AER, FAR and FRR rate were 8.17%, 2.5%, and 13.84%, respectively. Verification accuracy was 91.83%. A symbolic picture of a bi-interval valued of the signature was suggested. For binarization global threshold depends on the histogram that was implemented. For features, the comparative distance and positioning geometric centroid were utilized. Data set of MCYT-75 signature was used. 5, 7, and 9 genuine signatures of each person were considered for a set of training whereas others were used for testing. With the use of various centroid, for conducting verification, the algorithm of Max and Mean was used. The test signature was accepted to be genuine else forger if the count of acceptance of test signature was more than the threshold (predefined). The lower AER (AER = 18.26 for "MAX" fusion and AER = 17.33 for "MEAN" fusion) than the methods which straightly use any features of positioning or distance features.

For the verification of a signature, the graph matching problem was shown by Bibi et al. (2020). With the use of a hp ScanJet 3400C scanner, as binary images,75 genuine signatures were scanned. By employing methods of image processing, preprocessing of signature images was conducted. EER of 26.7% and 5.7% on skilled and random forgeries was attained by them. Through pixels intensity levels, the handwritten signature was discussed by Shah et al. (2016). For experimentation of offline signatures, a dataset of MCYT was used. Binarization was implemented and the images were resized and cropped to 120x120. At that time for better extraction of features, DWT was implemented. With the help of K-nearest neighbor naïve, decision tree, and Bayes tree, organization and verification were conducted. In the set of testing, 25% sets of data and in the set of training, 75% dataset was used by them. For 10 fold cross-validation was approved for simplification of outcomes. Between the three classifiers, k- the nearest neighbor had presented an 88.12% accurateness value (with forgeries) and more than 91.91% accurateness value (without forgeries). From offline signatures, the distinguishing features were extracted by Naamah et al. (2014). For experimentation, the Delphi programming language was considered. In preprocessing, normalization, thinning, Binarization, and other morphological processes were conducted. The signature was classified into areas. For the calculation of gravity center, for approximation of features, the method of COG was used. Through fuzzy logic, non-weighted, a directed graph with 64 nodes was formed. By algorithm of a dynamic classification, among trained and test signature, the ideal comparison was done. To investigate forgery, on a graph, EER was designed.

The work on Japanese katakana character, alphabet dataset taken from ETL-1 AIST database, handwritten character images of special characters, and numbers, learned from 1445 writers by Qasim et al. (2019). To change the images into binary, Otsu's global threshold was used. To obtain the skeleton, the algorithm of Zhang Suen was used then to take out curves from the character's skeleton images, connected component analysis was implemented.

The curves were shown in the string also from the curve feature, the graph string structure was labeled, taken out, and stored as information. For organizing data, approximate subgraph matching and string edit distance was considered. 10% of each label's image used the process of training and on 1000 and 2600 sets, the test was conducted in the technique of approximate subgraph matching.

On the entire data sets 8826 and 22647, the technique of string edit distance, the test was conducted, for every tag, each set of data includes a different images sets.

The accurateness was decreased from 82.4% to 77.7% in the outcomes of the 10% training data set. When the feature-length of the string curve was set as 8 and for training, 5% data was used, accurateness was improved from 77.7% to 83.6% (Al-Juboori, 2017).

5.4.2. Online Signature Verification

Like in verification systems of offline signature, in the field of verification systems of online signature, there are numerous methods. A few of the important works are described in the following segments.

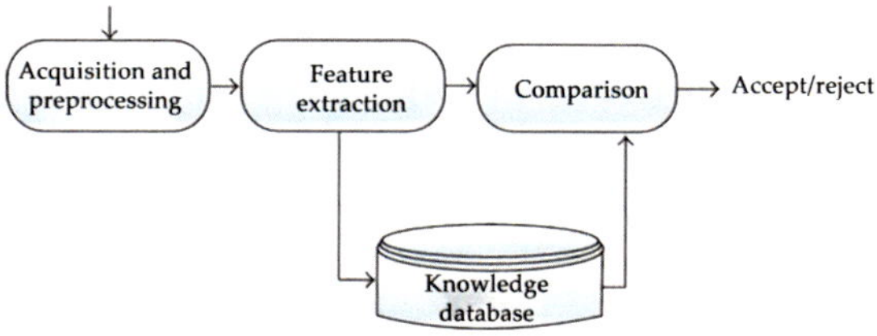

Figure 5.6. Block diagram of verification system of online signature.

Source: *https://www.researchgate.net/figure/Online-signature–verification-system-schema_fig1_264312002.*

5.4.2.1. Matching Technique

The procedure of comparing template/figure in which test signatures are compared with pre-stored templates in the database. In the verification system of online signature, the frequently used algorithm for this objective is DTW by researchers as mentioned here. On MCYT100 and SCV2004, the experiment was performed by Al-Hmouz et al. (2018). By the use of probabilistic dynamic time wrapping (PDTW), dynamic features were extracted and classified by them. In case 1, the stated lowest ERR on a dataset of the MCYT100 was 0.012% on the dataset of SCV2004 it was 0.002%. By using the method of dynamic time wrapping, the system was tested by Durrani et al. (2017). From the dataset of Standardized Japanese handwritten, the rate of the pen up and down and X, Y coordinates of signature were taken out. Without and with data of downsampling, testing was conducted. 27.35% FAR, 15.18% FRR, and attained accurateness of 79.80% was stated correspondingly by them.

On a database including 10 users, each having 10 forged and 10 genuine samples, the work was done by Borse and Patil (2017). From the image of input, data of Pen position was extracted. Through the algorithm of DTW, the normalization, pre-processing, and classification of data was done. The average accurateness rate of 90.4% was attained. From 14 persons having 30 forged and 10 genuine samples, a database of 560 signatures was developed by Fang et al. (2017). Firstly, the pre-processing of images of signatures occurred. Through frequency and time area features, the alteration of the algorithm of T-DTW-FFT fusion was occurred to get outdone performance. With a usual matching time of just 24 ms, 1.90% FAR and 2.86% FRR was stated. The working on the SVC2004 database and by utilizing the method of a distance, the signature was confirmed by Van et al. (2007). For testing, 20 forged and 15 genuine samples were used whereas for training, 5 genuine signatures were considered. eight features such as Timestamp, X, Y position, Azimuth, Pen status, Pressure, and Altitude were extracted from them. By employing the fast DTW method, the signature was confirmed by them. For methods of Average Permutative Difference Signature and Average Reference Signature, 1.84%, 1.36% FAR, and 0.31%, 0.51% FRR was reached by them.

MCYT database was utilized by Sundaram & Sharma (2016). The features such as Pressure, Spatial coordinates, and inclination, and azimuth angle were extracted by them. By utilizing Gaussian Mixture Model, statistical features were considered. Path covering verification, derivatives were calculated. To improve the verification, the scheme of Dynamic Time Warping matching was used. For the general threshold, an EER of 3.05% was attained. Through MCYT-100 datasets SVC2004 Task 2, the cost matrix of dynamic time wrapping was used by the same author for verification. For the test, 25 skilled and 20 forged signatures were considered whereas from MCYT-100, 5 genuine signatures were arbitrarily nominated. From SVC-2004, per 39 users, 1 genuine signature was nominated. The extraction of dynamic features such as attributes' time sequence was done and for the test signature matching, DTW was used and cost matrix and reference signature from them was formed. In the path of wrapping, the extraction of cost matrix cells was conducted from DTW. Alignment was considered (Sharma & Sundaram, 2016). Through numerical equations, the calculation of normalized average displacement and distortion was conducted. For validation, a fused core was determined.

From 42 subjects, the experiment was performed on 1680 forged and 1680 authentic signatures by Shin et al. (2017). All of the signatures had kanji Characters. From handwritten multi-stroke, intra-stroke, and inter-stroke information was extracted and the signatures were processed and normalized by them. Corresponding referenced signatures with Intra-stroke and Inter-stroke information, the algorithm of DP matching was used. The decreased FAR of 4.06% to 2.01% and FRR of 4.14% to 1.54%, was obtained by them respectively. On the SigComp2011 database, the work was done by Griechisch et al. (2014). By deducting and fluctuating x, y coordinates from original images, pre-processing was completed. The value of velocity was distinguished as features value. For determining the highest difference between the functions of cumulative distribution, the Kolmogo ROV-Smirnov test was implemented. For estimating data and training, the calculation of KS-distance of each respective reference signature was done. FRR and FAR of 8.02% and 7.86% was presented and for assessment, DTW was used by them. The accuracy and EER are almost 0.50% correspondingly.

On a database of 50 skilled and 50 genuine forgery samples, Impedovo & Pirlo (2008) did work. To get details of stroke-oriented, firstly the pre-processing of data and the implementation of the method of dynamic segmentation was conducted. In the area of velocity, displacement, and acceleration, the estimation of the stability and the examination of stroke features of test samples were performed. By weighted or simple averaging, the combination of outcomes of stroke verification was occurred to conduct verification and for matching points, consideration of DTW occurred. Experimentation on the SUSIG visual sub-corpus and MCYT online signature corpus was performed by Fischer et al. (2015). The features of pressure derivative, position, acceleration, and pressure were extracted. Through DTW, the matching was conducted and for evaluation, score normalization was used by them. For arbitrary forgeries, the average EER among 0.70%, 7.84% and for skilled forgeries, the average EER was 2.97% and 13.79% stated.

SVC 2004 and MCYT-100 were used in Sharma & Sundaram's (2016) research. Angle-based characteristics, as well as the X, Y coordinates and stress between successive locations, were retrieved. To standardize characteristics, we used zero mean and unit variance and zero mean. They used DTW to verify code vectors and conducted verification by matching them. On the SVC 2004 and MCYT-100 datasets, they attained FAR of 1.83% and 0.78% respectively. SUSIG and SVC2004 datasets were proposed by Ansari et al. (2014). On the samples, we conducted min-max normalization and segmentation. Each section was subjected to dynamic and form analysis. Each section was analyzed for its dynamic and form characteristics. Fuzzy modeling with the mixture of TS and Mamdani methods was used to create specific ranking points for each part.

To identify the shortest distance of two specimens, DTW was employed. On SVC2004 and SUSIG, they obtained EER of 2.46% and 1.3%, respectively, dependent upon this threshold level set by the user. Researchers used the SVC2004 data for their experiments. They divided the data into five real samples for retraining and 15 real, 20 fake signs for analysis. They used the DFT technique to extract features and the Fast DTW approach to match them. For ARS, they recorded FAR, FRR of 1.84%, 0.3% with an averaged ideal value of 0.527, while for APDS, they recorded FAR, FRR of 1.3%, 0.51% with an averaged optimum of 0.866. Dynamically retrieved stable characteristics were used by Song et al. (2016). For extracting spectrum data from characteristics, they used optimum mother wavelets and WP reconstructing. For accurate feature matching, modified Dynamic Time Wrapping (DTW) was used. On the SVC2004 task2 database, the suggested approach yielded an EER of 2.89%. Lopez-Garcia et al. (1993), developed an embedded system approach on the MCYT-100. The Gaussian Mixture Model was used to preprocess the signs and extract the features.

DTW was used to align the processed signature with the template that had been previously saved. EER was 2.74% for each of them. Using the Kinematic Theory of fast human motions and its related Sigma Log-Normal model, Gomez-Barrero et al. (2015) enhanced the performance of digital signature verification. On the BiosecurID multi-modal databases, DTW was utilized for verification and has a higher potential for skillful fake analysis. EER was improved by 36%, according to them. Rashidi et al. (2012) proposed an active approach for the match that used SVC2004 and the SUSIG dataset, as well as DCT and the parzen window classifier. Using DCT, they were able to extract characteristics such as the pen's location, angle, velocity, and pressure. For most optimum features subsets, the forward feature selection method was employed. The EER of the recommended approach was 3.61%, 2.04%, and 1.49% for the SVC2004 Task1and2, Task2, and SUSIG datasets, respectively. Online signature authentication utilizing the C++ computer language and 20 distinct features were given by Taherzadeh et al. (2011). Normalization, re-sampling time, and smoothing were used to pre-process the images. The DTW algorithm was used to match global and local characteristics. The suggested system's best AER was 15%, which was attained by combining five characteristics. Al-Hmouz et al. (2018) developed a novel approach of probabilistic dynamic time warping (PDTW) verification methodologies utilizing SCV2004 and MCYT100. The potential scores of signature pictures were assessed using relative distance after they were split into distinct segments. By connecting these segments, the Bayes rule was applied to arrive at conclusive findings. PDTW had a minimal improvement of 65% when compared to the DTW method's findings. For their tests, Tahir et al. (2016) employed the Japanese online exam set from ICDAR2013. Signature samples were analyzed and processed before being used. They used samples to extract x, y coordinates, as well as pen ups and downs locations. The referred and questioned signatures were matched using Euclidean distance and DTW. The suggested system has a segmentation accuracy of 78.14% and a non-segmentation accuracy of 78.57%.

5.4.2.2. Distance Measurements

In their paper, Gabor filter was used to extract characteristics without and with a time tag. Verification was done using the Euclidean distance approach. They reported an EER of percent without timestamp and a timestamp increase of up to 95%. Alizadeh et al. (2010) proposed an improved verification technique based on optimum threshold selection. From a digitizer plane, 62 parametric characteristics were derived. For comparison, the traditional weight Euclidean distance was used. For training, they utilized 30 real and 10 counterfeit samples, with the remaining samples to be used for the test. For 30 real and 10 counterfeit samples in the training set, they attained FRR and FAR of 0.67% and 1.33%, respectively, and 2.5% and 3% for the training dataset.

On the SUSIG dataset, Kaur & Kansa (2017) used a Hadamard transform-based method. The pictures of signatures were scanned and pre-processed. To build the Hadamard matrix, features have been extracted. The Euclidean and Manhattan distances were used to compare and verify the sizes of the creatures. The proposed method's reported EER was 0.05% using Manhattan 0.03% and 0.025 using Euclidean. Philip & Bharadi (2016) used Azure Blob storage to extract characteristics such as X, Y coordinates, pressure values, and Weber Local Descriptor (WLD) for digital signature verification. They used K-nearest neighbor classification to classify feature vectors supplied to blob storage. The suggested system has a PI of 92.50% and a CCR of 94.25%.

Nisha and Deepesh (2017) generated a collection of 80 signature verification from 15 people. They used Z-score normalization to normalize scanned data. The SFFS method was used to obtain 28 feature vectors and 7 local features. For global feature verification, they utilized Mahalanobis distance, and for the temporal function of signatures, they used DTW. KNN was used to do the classification. The method outperformed previous systems by 38.5% EER and 13.0% EER for competent forgeries. Cho and Jung (2017) proposed a new pseudo-ink-based approach for utilizing signature verification. For the trials, they employed 105 reference pictures and 2625 testing data. The value of the pen pressure, the tilting angle of the pen, and the theta angle of the pen were all retrieved and classified. And use the threshold value of Pen pressure values, the system displayed FAR, FRR of (0.2381%, 0.2381%) and (4.7619, 4.7619) for authentic signings and FAR of 4.7619% for signed documents (0.55 to 0.65). Hafs et al. (2016) proposed a technique based on the SVC2004 and MYCT-100 datasets. In preprocessing, they utilized a Gaussian filter and a distance filter and then used the EMD method to extract dynamic characteristics. For comparison, the Euclidean distance was used to obtain the similarity score.

For SVC2004 task 1 and 2.23% for MYCT-100, the suggested system's EER was 1.83% and 2.23%, respectively. Nguyen et al. (2011) used the GPDS-960 signature corpus, which has 160 signature sets. For picture conversion, they employed Otsu's thresholding method. The data were divided into two groups: 12 real forgeries and 400 random forgeries for testing, and 12 real forgeries and 15 forgeries for testing. The gradient feature and the modified direction feature were extracted. For classification, the Squared Mahalanobis distance and SVM were employed. The system generated an AER and EER of 16.52% and 16.77% for the gradient feature when using Mahalanobis distance, and an AER and EER of 15.03% and 15.11% when using SVM. The FAR and FRR were respectively 16.54% and 13.51%. the MCYT-100 dataset in experiments was used (Nilchiyan & Yusof, 2013). They utilized five real samples for learning, the rest of the samples for testing, and 25 forgeries for training. To connect data sets, dynamic time warping was chosen. For the verification job, Mahalanobis distance was used to represent position and pressure velocity as feature vectors. The MD-based technique outperformed the ED-based method, according to the system.

5.4.2.3. Neural Networks

Iranmanesh et al. (2013) utilized the MLP neural network to classify data from 200 users in the SIGMA dataset. For training and testing, a total of 4000 signature samples were employed. To extract features, Pearson's correlation coefficient was chosen. The suggested method has a 21.35% success rate, 13.81% FAR, and 82.42% accuracy. Using normalizing and resampling pictures, Fahmy (2010) retrieved Discrete Wavelet Transform (DWT) characteristics such as pen location and pen movement angle. They tested 20 authentic signatures against 20 competent forgery signatures on the SVC2004 dataset. For the classification challenge, six neural networks were employed. In the event of authentic signatures, access to the system has a 95% rate of success., Jain & Gangrade (2013) used a dataset of 300 signature samples taken from six persons. 180 samples were utilized for training and 120 samples have been used for testing.

Chain-Code, Angle, and energy density methods were used to extract features from pre-processed samples. The categorization, which was done with a neural network, was accurate to the tune of 70% to 80%. Malallah et al. (2015) presented a method for utilizing the SIGMA database based on artificial neural networks (ANN). A sample of 200 signatures was gathered, with 10 authentic and 10 fake samples were chosen for testing. Using PCA, they retrieved characteristics such as horizontally and vertically to time-series signals, as well as pen pressure. By upsampling and downsampling and Upsampling the samples, they were normalized. FAR and FRR of (5.50 and 8.75%, respectively, were obtained using this technique. In their study, Nilchiyan and Yusof employed the SVC 2004 database, which had 15 real and 15 counterfeit signatures. Wavelet transformation is used to obtain statistical characteristics. For improved quality and cost savings, the B-spline function was employed. Mul was next. Malallah et al. (2015) proposed an artificial neural network-based approach for using the SIGMA dataset (ANN).

A total of 200 signatures were collected, with t10 genuine and 10 fake samples being tested. They recovered features like vertically and horizontally to time series data, as well as pen pressure, using PCA. The samples were normalized by upsampling, downsampling, and upsampling. This method yielded a FAR and FRR of 5.50 and 8.75%, respectively. Nilchiyan & Yusof (2013) used the SVC 2004 database to conduct their research, which included 15 actual and 15 fake signatures. To get statistical characteristics, the wavelet transformation is performed. The B-spline function was utilized to improve quality and cut costs. The signatures were then classified using a multi-perceptron neural network, yielding an overall EER of 3.5%.

Xu et al. (2020) developed a novel technique for signatures identification based on BP neural networks. Eight volunteers gathered a total of 400 signatures. As a feature set, a total of 49 features were computed. The suggested system's overall acquired classification accuracy was around 95%. The Pearson's correlation coefficient technique for extracting features of digital signatures is introduced by Iranmanesh et al. (2013). For the testing and training process, they chose 5 non-skilled and skilled forgeries, 10 real, from both the SIGMA dataset. Using a Deep Network, 9 distinct characteristics were retrieved and verified. The suggested approach's FAR, FRR, and accuracy were 21.25, 13.81, and 82.42%, respectively. Chadha et al. (2013) used a dataset of 700 samples and collected 10 samples across 70 persons to create their study. In the beginning, pictures were processed using rotation-translation-scaling. DCT was used to extract the features, and RBFN was used to determine the signature. For roughly 200 samples, the system obtained an 80% identification rate.

5.4.2.4. Hidden Markov Model

Using the MCYT signature database, Nanni et al. (2010) introduced a novel verification method based on a mix of locally, worldwide and regional matches. As matching approaches, the HMM, DTW, and a Linear Programming Descriptor were used. The EER was 4.51% using Bio-Convolving and Bio-Hashing methods with a templates protection strategy, which is quite low. From the other side, the suggested matching technique produces a 3% more effective outcome.

5.4.2.5. Support Vector Machine

Durrani et al. (2016) used an SVM classifier to perform an analysis using ICDAR 2013 signature pictures. For instruction, they used 396 fabricated samples, whereas, for testing, they used 42 real and 36 faked signs per author. 1055 was the obtained threshold value. The database was used to extract and store and extract the dynamic set of features. The system has an EER of 20% and reliability of 75%.

Khoh et al. (2014) researched the SVC2004 Task 2 and Task 1 database. Pre-processed and scanned pictures were used. To extract feature sets from every signature signal, DFT and DTW were employed. For comparing the features of the exam and reference signatures, they employed distance measure and EED. In Task 1, EER was 6.55% for competent signatures and 1.29% for accidental forgeries. In Task 2, EER was 7.17% for competent signatures and 1.15% for random signatures.

Griechisch et al. (2014) established a novel method for certification by comparing the AccSigDb2011) and (GyroSigDb2012) databases, which provide local accelerating and angle information. For feature computation, they utilized Legendre approximations, while SVM was used for categorization. In AccSigDb2011, the overall result was approximately 85%, but in GyroSigDb2012, the reliability was lower. GyroSigDb2012 had a 50% accuracy when utilizing multiclass classification percent, but AccSigDb2011 had a 35% accuracy when the number of the Lagrange polynomials was smaller than 20.

Rosso et al. (2016) investigated the MCYT data set, which is utilized for quasi-offline signature. The signatures coordinate information is considered quasi-offline. The pictures were scanned, and the time series of both horizontally and vertically writing operations were extracted using interpolation. For each coordinate, a 6-dimensional vector was recovered and distribution was constructed. SVM was used for categorization. With

5, 10 training sets, the EER of the suggested approach was 0.19 and 0.17%, respectively. Manjunatha et al. (2016) conducted experiments on MCYT-100 and MCYT-330. A total of 100 global characteristics were taken into account. For classification, six different statistics classifiers were utilized, including Naive Bayesian, Closest Neighbor, SVM, Principal, Probabilistic Neural Network, Component Analysis, and Logistic Regression. For the training phase, they utilized 5 to 20 authentic signatures per writer. FRR, FAR, and EER were 3.83, 0, and 1.92%, respectively. A smartphone was used to collect 2,940 digital signatures from 42 respondents. Using Python, 57 features have been extracted, and 57-dimensional extracted features with 50 positives and 20 negative samples extracted features were produced.

The categorization was done using SVM, Linear Regression, AdaBoost, Random Forest, and AdaBoost. Using AdaBoost, the system obtained a failure rate of 2.4%. Fayyaz et al. (2015) experimented with the SVC2004 dataset. PCA was used to scan the images, standardize them, and reduce their size. The unsupervised classification was used to extract features. SVM with RBF & KNN classifiers was used for testing with signatures. The model obtained EER of 2.15 and 99.11%.

5.4.2.6. Structural Approaches

Graphs, trees, and strings are employed in structural methods to display example patterns. With the network's patterns restored, the symbolic representation would be calculated.

Wang et al. (2013) developed a technique based on the SUSIG dataset. First, signs were represented by a series of graphs, and later graph matching algorithms were used to compute the distance measure between graphs, which was used to determine how similar the charts were. User-dependent classifiers are used for categorization. Shah et al. (188) used three classification methods in their comparative analysis: decision tree (J48), Naive Bayes (NB tree), and KNN for confirmation. They employed 350 signatures from the ATVS-SSig databases. The DWT was utilized to extract features. Using decision trees, 99.90% accuracy was attained, 99.82% using naive Bayes, and 98.11% using distance-based grand neighbor classifiers.

Pirlo et al. (2015) researched the SUSIG database, using 10 real forgeries and 5 expert forgeries for training and assessment. Prepossessing was done using linear normalizing and linear interpolation algorithms. Samples were used to extract displacement, velocity, acceleration, velocity, and pressure function characteristics. DTW was utilized to pick prototypes and identify the wrapping method. After that, we created the signer's stability model and used a decision tree to verify it. Signature variation was calculated using the standard deviation technique, and efficiency was improved using the probabilistic acceptance model. FRR and FAR were 2.15 and 2.10%, respectively (Patil & Hegadi, 2013).

5.5. TRENDS AND CHALLENGES OF A SIGNATURE VERIFICATION SYSTEM

In contrast to other biometric technologies, handwritten signatures may be openly disputed at any moment. Because of the diversity of problems it presents, despite considerable study signature verification stays open to the scientific community. We outline some of the significant problems of handwritten signature recognition methods in this part, which pave the way for future study in this field. One of the primary issues inside the offline signature creation process, starting with the picture capture phase, is the loss of dynamic information. Scanning the database showing signature information captures the offline signatures. Dynamic information such as pen location, speed, and velocity of writing is frequently lost during scanning, resulting in a poor recognition rate and making the problem difficult to solve (Philip & Bharadi, 2016).

This problem is solved with online signature verification systems, which collect signature data via online devices such as digital pens and tablets, PDAs, graphic tablets, or even digital gloves. Although online systems give dynamic information, the availability of internet devices remains a critical element. Furthermore, internet devices should adhere to a set of standards to provide the highest level of accuracy, validity, and systems legality. These standards differ by nation; for example, the electronics Identity verification, Authentication, and Trusted Services (eIDAS) standards are used in Europe (Prabhakare et al., 2003).

REFERENCES

1. Abikoye, O. C., Mabayoje, M. A., & Ajibade, R. (2011). Offline signature recognition & verification using neural network. *International Journal of Computer Applications*, *35*(2), 44–51.
2. Ahmad, R., Naz, S., Afzal, M. Z., Amin, S. H., & Breuel, T. (2015). Robust optical recognition of cursive Pashto script using scale, rotation and location invariant approach. *PloS One*, *10*(9), e0133648.
3. Ahmed, S. B., Naz, S., Razzak, M. I., & Yusof, R. (2019). Arabic cursive text recognition from natural scene images. *Applied Sciences*, *9*(2), 236.
4. Ahmed, S. B., Naz, S., Razzak, M. I., & Yusof, R. B. (2019). A novel dataset for English-Arabic scene text recognition (EASTR)-42K and its evaluation using invariant feature extraction on detected extremal regions. *IEEE Access*, *7*, 19801–19820.
5. Al-Hmouz, R., Pedrycz, W., Daqrouq, K., & Morfeq, A. (2018). Development of multimodal biometric systems with three-way and fuzzy set-based decision mechanisms. *International Journal of Fuzzy Systems*, *20*(1), 128–140.
6. Al-Hmouz, R., Pedrycz, W., Daqrouq, K., Morfeq, A., & Al-Hmouz, A. (2019). Quantifying dynamic time warping distance using probabilistic model in verification of dynamic signatures. *Soft Computing*, *23*(2), 407–418.
7. Alizadeh, A., Alizadeh, T., & Daei, Z. (2010, March). Optimal threshold selection for online verification of signature. In *Proceedings of the International Multiconference of Engineers and Computer Scientists*, 1, 17–19.
8. Al-Juboori, S. S. (2017). Signature verification based on moments technique. *Ibn AL-Haitham Journal For Pure and Applied Science*, *26*(2), 385–395.
9. Al-Mayyan, W., Own, H. S., & Zedan, H. (2011). Rough set approach to online signature identification. *Digital Signal Processing*, *21*(3), 477–485.
10. Al-Omari, Y. M., Abdullah, S. N. H. S., & Omar, K. (2011, June). State-of-the-art in offline signature verification system. In *2011 International Conference on Pattern Analysis and Intelligence Robotics*, 1, 59–64.
11. Ansari, A. Q., Hanmandlu, M., Kour, J., & Singh, A. K. (2014). Online signature verification using segment-level fuzzy modelling. *IET Biometrics*, *3*(3), 113–127.

12. Arunalatha, J. S., Prashanth, C. R., Tejaswi, V., Shaila, K., Raja, K. B., Anvekar, D., ... & Pawan, K. S. (2015). OSPCV: Off-line Signature Verification using Principal Component Variances. *IOSR Journal of Computer Engineering (IOSR-JCE)*, *17*(2015), 08–23.
13. Azmi, A. N., Nasien, D., & Omar, F. S. (2017). Biometric signature verification system based on freeman chain code and k-nearest neighbor. *Multimedia Tools and Applications*, *76*(14), 15341–15355.
14. Bashir, S., Sofi, S., Aggarwal, S., & Singhal, S. (2015). Unimodal & multimodal biometric recognition techniques a survey. *International Journal of Computer Science and Network*, *4*(1), 148–155.
15. Bharadi, V. A., & Kekre, H. B. (2010). Off-line signature recognition systems. *International Journal of Computer Applications*, *1*(27), 48–56.
16. Bhunia, A. K., Alaei, A., & Roy, P. P. (2019). Signature verification approach using fusion of hybrid texture features. *Neural Computing and Applications*, *31*(12), 8737–8748.
17. Bibi, K., Naz, S., & Rehman, A. (2020). Biometric signature authentication using machine learning techniques: Current trends, challenges and opportunities. *Multimedia Tools and Applications*, *79*(1), 289–340.
18. Chadha, A., Satam, N., & Wali, V. (2013). Biometric signature processing & recognition using radial basis function network. *arXiv preprint arXiv:1311.1694*.
19. Cho, Y. O., & Jung, J. W. (2017). Online signature recognition based on pseudo-inked signature image template. *International Journal of Humanoid Robotics*, *14*(02), 1750016.
20. Coetzer, J., Herbst, B. M., & du Preez, J. A. (2004). Offline signature verification using the discrete radon transform and a hidden Markov model. *EURASIP Journal on Advances in Signal Processing*, *2004*(4), 1–13.
21. Deore, M. R., & Handore, S. M. (2015, May). A survey on offline signature recognition and verification schemes. In *2015 International Conference on Industrial Instrumentation and Control (ICIC)*, 3, 165–169.
22. Durrani, M. Y., Ali, A., Mustafa, A., & Khalid, S. (2016). Signature verification through chebyshev polynomials a novel method. *Journal Applied Environmental and Biological Science*, *6*(3), 159–165.

23. Durrani, M. Y., Khan, S., Khan, M. T., Ali, A., Mustafa, A., & Khalid, S. (2017). Signature Identification Through Decision Envelope a Novel Approach. *Journal of Computational and Theoretical Nanoscience*, *14*(2), 1204–1209.
24. Fahmy, M. M. (2010). Online handwritten signature verification system based on DWT features extraction and neural network classification. *Ain Shams Engineering Journal*, *1*(1), 59–70.
25. Fang, Y., Kang, W., Wu, Q., & Tang, L. (2017). A novel video-based system for in-air signature verification. *Computers & Electrical Engineering*, *57*, 1–14.
26. Fayyaz, M., Saffar, M. H., Sabokrou, M., Hoseini, M., & Fathy, M. (2015, March). Online signature verification based on feature representation. In *2015 The International Symposium on Artificial Intelligence and Signal Processing (AISP)*, 4, 211–216.
27. Fischer, A., Diaz, M., Plamondon, R., & Ferrer, M. A. (2015, August). Robust score normalization for DTW-based on-line signature verification. In *2015 13th International Conference on Document Analysis and Recognition, 4,* 241–245.
28. Garhawal, S., & Shukla, N. (2013). Surf based design and implementation for handwritten signature verification. *International Journal of Advanced Research in Computer Science and Software Engineering (IJARCSSE)*, *3*(8), 3–10.
29. Gomez-Barrero, M., Galbally, J., Fierrez, J., Ortega-Garcia, J., & Plamondon, R. (2015, May). Enhanced on-line signature verification based on skilled forgery detection using sigma-lognormal features. In *2015 International Conference on Biometrics, 4(1),* 501–506.
30. Griechisch, E., Malik, M. I., & Liwicki, M. (2014, September). Online signature verification based on kolmogorov-smirnov distribution distance. In *2014 14th International Conference on Frontiers in Handwriting Recognition*, 4(1), 738–742.
31. Griechisch, E., Malk, M. I., & Liwicki, M. (2013, August). Online signature analysis based on accelerometric and gyroscopic pens and legendre series. In *2013 12th International Conference on Document Analysis and Recognition*, 5(2), 374–378.
32. Griechisch, W. H., Ong, T. S., Pang, Y. H., & Teoh, A. B. J. (2014). Score level fusion approach in dynamic signature verification based on hybrid wavelet-Fourier transform. *Security and Communication Networks*, *7*(7), 1067–1078.

33. Hafemann, L. G., Oliveira, L. S., & Sabourin, R. (2018). Fixed-sized representation learning from offline handwritten signatures of different sizes. *International Journal on Document Analysis and Recognition (IJDAR)*, *21*(3), 219–232.
34. Hafemann, L. G., Sabourin, R., & Oliveira, L. S. (2017). Learning features for offline handwritten signature verification using deep convolutional neural networks. *Pattern Recognition*, *70*, 163–176.
35. Hafemann, L. G., Sabourin, R., & Oliveira, L. S. (2019). Characterizing and evaluating adversarial examples for Offline Handwritten Signature Verification. *IEEE Transactions on Information Forensics and Security*, *14*(8), 2153–2166.
36. Hafs, T., Bennacer, L., Boughazi, M., & Nait-Ali, A. (2016). Empirical mode decomposition for online handwritten signature verification. *IET Biometrics*, *5*(3), 190–199.
37. Hanmandlu, M., Yusof, M. H. M., & Madasu, V. K. (2005). Off-line signature verification and forgery detection using fuzzy modeling. *Pattern Recognition*, *38*(3), 341–356.
38. Herbst, N. M., & Liu, C. N. (1977). Automatic signature verification based on accelerometry. *IBM Journal of Research and Development*, *21*(3), 245–253.
39. Huang, K., & Yan, H. (1997). Off-line signature verification based on geometric feature extraction and neural network classification. *Pattern Recognition*, *30*(1), 9–17.
40. Impedovo, D., & Pirlo, G. (2008). Automatic signature verification: The state of the art. *IEEE Transactions on Systems, Man, and Cybernetics, Part C (Applications and Reviews)*, *38*(5), 609–635.
41. Iranmanesh, V., Ahmad, S. M. S., Adnan, W. A. W., Malallah, F. L., & Yussof, S. (2013, December). Online signature verification using neural network and pearson correlation features. In *2013 IEEE Conference on Open Systems, 4,* 18–21.
42. Jain, P., & Gangrade, J. (2013). Online Signature Verification using Energy Angle and Directional Gradient Feature with Neural Network. *International Journal of Innovative Research in Science Engineering and Technology*, *2*(9, 2–10.
43. Kalenova, D. (2003). Personal authentication using signature recognition. *Department of Information Technology, Laboratory of Information Processing, Lappeenranta University of Technology*, *3*, 1–15.

44. Kaur, H., & Kansal, E. R. (2017). Distance based online signature verification with enhanced security. *International Journal of Engineering Development and Research (IJEDR)*, *5*(2), 1703–1710.
45. Kaur, R., & Choudhary, M. P. (2015). Handwritten signature verification based on surf features using HMM. *International Journal of Computer Science Trends and Technology (IJCST)*, *3*(1), 187–195.
46. Kennard, D. J., Barrett, W. A., & Sederberg, T. W. (2012, November). Offline signature verification and forgery detection using a 2-D geometric warping approach. In *Proceedings of the 21st International Conference on Pattern Recognition, 43,* 3733–3736.
47. Kiani, V., Pourreza, R., & Pourreza, H. R. (2009). Offline signature verification using local radon transform and support vector machines. *International Journal of Image Processing*, *3*(5), 184–194.
48. Kruthi, C., & Shet, D. C. (2014, January). Offline signature verification using support vector machine. In *2014 Fifth International Conference on Signal and Image Processing*, 5, 3–8.
49. Kumar, D. A., & Dhandapani, S. (2016). A novel bank check signature verification model using concentric circle masking features and its performance analysis over various neural network training functions. *Indian Journal of Science and Technology*, *9*(31), 2–10.
50. Kumar, M. M., & Puhan, N. B. (2014). Off-line signature verification: upper and lower envelope shape analysis using chord moments. *IET Biometrics*, *3*(4), 347–354.
51. Kumar, P., Singh, S., Garg, A., & Prabhat, N. (2013). Hand written signature recognition & verification using neural network. *International Journal of Advanced Research in Computer Science and Software Engineering*, *3*(3), 5–10.
52. Li, N., Liu, J., Li, Q., Luo, X., & Duan, J. (2016, January). Online signature verification based on biometric features. In *2016 49th Hawaii International Conference on System Sciences, 412,* 5527–5534.
53. Liwicki, M., Malik, M. I., Van Den Heuvel, C. E., Chen, X., Berger, C., Stoel, R., ... & Found, B. (2011, September). Signature verification competition for online and offline skilled forgeries (sigcomp2011). In *2011 International Conference on Document Analysis and Recognition*, 321, 1480–1484.
54. Lopez-Garcia, J. A., Ramis, C., Nicolau, M. C., Alemany, G., Planas, B., & Rial, R. (1993). Histaminergic drugs in the rat caudate nucleus:

effects on learned helplessness. *Pharmacology Biochemistry and Behavior*, *45*(2), 275–282.

55. Madabusi, S., Srinivas, V., Bhaskaran, S., & Balasubramanian, M. (2005, March). On-line and off-line signature verification using relative slope algorithm. In *Proceedings of the 2005 IEEE International Workshop on Measurement Systems for Homeland Security, Contraband Detection and Personal Safety Workshop, 232,* 11–15.
56. Malallah, F. L., Ahmad, S. M. S., Adnan, W. A. W., Arigbabu, O. A., Iranmanesh, V., & Yussof, S. (2015). Online handwritten signature recognition by length normalization using up-sampling and down-sampling. *Int. J. Cyber Secur. Digit. Foren*, *4*, 302–13.
57. Manjunatha, K. S., Manjunath, S., Guru, D. S., & Somashekara, M. T. (2016). Online signature verification based on writer dependent features and classifiers. *Pattern Recognition Letters*, *80*, 129–136.
58. Maruyama, T. M., Oliveira, L. S., Britto, A. S., & Sabourin, R. (2020). Intrapersonal parameter optimization for offline handwritten signature augmentation. *IEEE Transactions on Information Forensics and Security*, *16*, 1335–1350.
59. Mehta, M., Choudhary, V., Das, R., & Khan, I. (2010, February). Offline signature verification and skilled forgery detection using HMM and sum graph features with ANN and knowledge based classifier. In *Second International Conference on Digital Image Processing*, 7546, 75462.
60. Moos, S., Marcolin, F., Tornincasa, S., Vezzetti, E., Violante, M. G., Fracastoro, G., ... & Padula, F. (2017). Cleft lip pathology diagnosis and foetal landmark extraction via 3D geometrical analysis. *International Journal on Interactive Design and Manufacturing (IJIDeM)*, *11*(1), 1–18.
61. Nagel, R. N., & Rosenfeld, A. (1973, October). Steps toward handwritten signature verification. In *Proc. First Int. Joint Conf. on Pattern Recognition, Washington, 21,* 59–66.
62. Nandi, G. C., Semwal, V. B., Raj, M., & Jindal, A. (2016, November). Modeling bipedal locomotion trajectories using hybrid automata. In *2016 IEEE Region 10 Conference, 321,* 1013–1018.
63. Nanni, L., Maiorana, E., Lumini, A., & Campisi, P. (2010). Combining local, regional and global matchers for a template protected on-line signature verification system. *Expert Systems with Applications*, *37*(5), 3676–3684.

64. Naseer, A., Rani, M., Naz, S., Razzak, M. I., Imran, M., & Xu, G. (2020). Refining Parkinson's neurological disorder identification through deep transfer learning. *Neural Computing and Applications*, *32*(3), 839–854.
65. Neamah, K., Mohamad, D., Saba, T., & Rehman, A. (2014). Discriminative features mining for offline handwritten signature verification. *3D Research*, *5*(1), 1–6.
66. Nguyen, V., & Blumenstein, M. (2011, September). An application of the 2d gaussian filter for enhancing feature extraction in off-line signature verification. In *2011 International Conference on Document Analysis and Recognition*, 23, 339–343.
67. Nguyen, V., Blumenstein, M., Muthukkumarasamy, V., & Leedham, G. (2007, September). Off-line signature verification using enhanced modified direction features in conjunction with neural classifiers and support vector machines. In *Ninth International Conference on Document Analysis and Recognition,* 2, 734–738.
68. Nilchiyan, M. R., & Yusof, R. B. (2013, December). Improved wavelet-based online signature verification scheme considering pen scenario information. In *2013 1st International Conference on Artificial Intelligence, Modelling and Simulation, 43,* 8–13.
69. Nisha, S., & Deepesh, A. (2017). Online biometric signature verification based on both local and global system. *International Research Journal of Engineering and Technology (IRJET)*, *4*(5), 2298–2300.
70. Okawa, M. (2018). From BoVW to VLAD with KAZE features: Offline signature verification considering cognitive processes of forensic experts. *Pattern Recognition Letters*, *113*, 75–82.
71. Pal, S., Alireza, A., Pal, U., & Blumenstein, M. (2011, December). Off-line signature identification using background and foreground information. In *2011 International Conference on Digital Image Computing: Techniques and Applications*, 21, 672–677.
72. Pal, S., Blumenstein, M., & Pal, U. (2011, February). Off-line signature verification systems: a survey. In *Proceedings of the International Conference & Workshop on Emerging Trends in Technology*, *53*, 652–657.
73. Pal, S., Chanda, S., Pal, U., Franke, K., & Blumenstein, M. (2012, November). Off-line signature verification using G-SURF. In *2012 12th International Conference on Intelligent Systems Design and Applications, 43,* 586–591.

74. Pal, S., Pal, U., & Blumenstein, M. (2017). An efficient signature verification method based on an interval symbolic representation and a fuzzy similarity measure. *IEEE Transactions on Information Forensics and Security*, *12*(10), 2360–2372.

75. Parodi, M., & Gómez, J. C. (2014). Legendre polynomials based feature extraction for online signature verification. Consistency analysis of feature combinations. *Pattern Recognition*, *47*(1), 128–140.

76. Parziale, A., Diaz, M., Ferrer, M. A., & Marcelli, A. (2019). Sm-dtw: Stability modulated dynamic time warping for signature verification. *Pattern Recognition Letters*, *121*, 113–122.

77. Patel Bhuminha, A., & Kumar, S. (2015). A survey on handwritten signature verification Techniques. *Insrumentation journal of Advance Research in computer Science and Management Studies*, *3*(1), 2–20.

78. Patil, P. G., & Hegadi, R. S. (2013). Offline handwritten signatures classification using wavelets and support vector machines. *International Journal of Engineering Science and Innovative Technology*, *2*(4), 573–79.

79. Patil, V. R., Borse, Y. I., Patil, R. R., & Patil, M. P. (2017). Online signature verification using dtw algorithm: A review. *International Journal*, *2*(7), 232–243.

80. Philip, J., & Bharadi, V. A. (2016). Signature verification SaaS implementation on microsoft azure cloud. *Procedia Computer Science*, *79*, 410–418.

81. Pirlo, G., Cuccovillo, V., Diaz-Cabrera, M., Impedovo, D., & Mignone, P. (2015). Multidomain verification of dynamic signatures using local stability analysis. *IEEE Transactions on Human-Machine Systems*, *45*(6), 805–810.

82. Prabhakar, S., Pankanti, S., & Jain, A. K. (2003). Biometric recognition: Security and privacy concerns. *IEEE security & privacy*, *1*(2), 33–42.

83. Prakash, H. N., & Guru, D. S. (2010). Offline signature verification: An approach based on score level fusion. *International Journal of Computer Applications*, *1*(1), 52–58.

84. Pushpalatha, K. N., Gautham, A. K., Shashikumar, D. R., ShivaKumar, K. B., & Das, R. (2013). Offline signature verification with random and skilled forgery detection using polar domain features and multi stage classification-regression model. *International Journal of Advanced Science and Technology*, *59*, 27–40.

85. Qasim, M., Amin, M., Akram, M. N., Omer, T., & Hussain, F. (2019). Forecasting Buffalo Population of Pakistan using Autoregressive Integrated Moving Average (ARIMA) Time Series Models: Forecasting Buffalo Population of Pakistan. *Proceedings of the Pakistan Academy of Sciences: A. Physical and Computational Sciences*, *56*(3), 27–36.
86. Randhawa, M. K., Sharma, A. K., & Sharma, R. K. (2012). Off-line signature verification based on Hu's moment invariants and zone features using support vector machine. *International Journal of Latest Trends in Engineering and Technology*, *1*(3), 16–23.
87. Randhawa, M. K., Sharma, A. K., & Sharma, R. K. (2013, February). Off-line signature verification with concentric squares and slope based features using support vector machines. In *2013 3rd IEEE International Advance Computing Conference, 323,* 600–604.
88. Rashidi, S., Fallah, A., & Towhidkhah, F. (2012). Feature extraction based DCT on dynamic signature verification. *Scientia Iranica*, *19*(6), 1810–1819.
89. Rosso, O. A., Ospina, R., & Frery, A. C. (2016). Classification and verification of handwritten signatures with time causal information theory quantifiers. *PloS one*, *11*(12), e0166868.
90. Serdouk, Y., Nemmour, H., & Chibani, Y. (2016). New off-line handwritten signature verification method based on artificial immune recognition system. *Expert Systems with Applications*, *51*, 186–194.
91. Shah, A. S., Khan, M. N. A., Subhan, F., Fayaz, M., & Shah, A. (2016). An offline signature verification technique using pixels intensity levels. *International Journal of Signal Processing, Image Processing and Pattern Recognition*, *9*(8), 205–222.
92. Sharif, M., Khan, M. A., Faisal, M., Yasmin, M., & Fernandes, S. L. (2020). A framework for offline signature verification system: Best features selection approach. *Pattern Recognition Letters*, *139*, 50–59.
93. Sharma, A., & Sundaram, S. (2016). A novel online signature verification system based on GMM features in a DTW framework. *IEEE Transactions on Information Forensics and Security*, *12*(3), 705–718.
94. Sharma, A., & Sundaram, S. (2016). An enhanced contextual DTW based system for online signature verification using vector quantization. *Pattern Recognition Letters*, *84*, 22–28.

95. Shekar, B. H., & Bharathi, R. K. (2011, June). Eigen-signature: A robust and an efficient offline signature verification algorithm. In *2011 International Conference on Recent Trends in Information Technology, 43,* 134–138.

96. Shekar, B. H., Pilar, B., & Kumar, D. S. (2019). Offline signature verification based on partial sum of second-order taylor series expansion. In *Data Analytics and Learning*, 43, 359–367.

97. Shin, J., Maruyama, K., & Kim, C. M. (2017). Signature verification based on inter-stroke and intra-stroke information. *ACM SIGAPP Applied Computing Review*, *17*(1), 26–34.

98. Shirazi, S. H., Umar, A. I., Haq, N., Naz, S., Razzak, M. I., & Zaib, A. (2018). Extreme learning machine based microscopic red blood cells classification. *Cluster Computing*, *21*(1), 691–701.

99. Song, X., Xia, X., & Luan, F. (2016). Online signature verification based on stable features extracted dynamically. *IEEE Transactions on Systems, Man, and Cybernetics: Systems*, *47*(10), 2663–2676.

100. Souza, V. L., Oliveira, A. L., Cruz, R. M., & Sabourin, R. (2020). A white-box analysis on the writer-independent dichotomy transformation applied to offline handwritten signature verification. *Expert Systems with Applications*, *154*, 113397.

101. Taherzadeh, G., Karimi, R., Ghobadi, A., & Beh, H. M. (2011, February). Evaluation of online signature verification features. In *13th International Conference on Advanced Communication Technology, 6,* 772–777.

102. Tahir, M., & Akram, M. U. (2015, November). Online signature verification using hybrid features. In *2015 Second International Conference on Information Security and Cyber Forensics, 7,* 11–16.

103. Tahir, M., Akram, M. U., & Idris, M. A. (2016, April). Online signature verification using segmented local features. In *2016 International Conference on Computing, Electronic and Electrical Engineering, 7,* 100–105.

104. Thakare, B., Deshmukh, H., & Mahalle, P. (2016). Handwritten signatures: An understanding. *International Journal of Computer Applications*, *139*(4), 21–26.

105. Van, B. L., Garcia-Salicetti, S., & Dorizzi, B. (2007). On using the Viterbi path along with HMM likelihood information for online signature verification. *IEEE Transactions on Systems, Man, and Cybernetics, Part B (Cybernetics)*, *37*(5), 1237–1247.

106. Vargas, F., Ferrer, M., Travieso, C., & Alonso, J. (2007, September). Off-line handwritten signature GPDS-960 corpus. In *Ninth International Conference on Document Analysis and Recognition,* 2, 764–768.

107. Vargas, J. F., Ferrer, M. A., Travieso, C. M., & Alonso, J. B. (2011). Off-line signature verification based on grey level information using texture features. *Pattern Recognition*, *44*(2), 375–385.

108. Xiong, J., Zhu, J., Wang, K., Wang, X., Ye, X., Liu, L., ... & Zhang, D. (2014). The temporal scaling of bacterioplankton composition: high turnover and predictability during shrimp cultivation. *Microbial Ecology*, *67*(2), 256–264.

109. Yang, X., Qiao, H., & Liu, Z. Y. (2015). Feature correspondence based on directed structural model matching. *Image and Vision Computing*, *33*, 57–67.

110. Zhang, L., Zhou, Y., Cheng, C., Cui, H., Cheng, L., Kong, P., ... & Cui, Y. (2015). Genomic analyses reveal mutational signatures and frequently altered genes in esophageal squamous cell carcinoma. *The American Journal of Human Genetics*, *96*(4), 597–611.

Chapter 6

Fingerprints Classification Using Machine Learning and Image Analysis

CONTENTS

6.1. INTRODUCTION

The structure that automatically recognizes the anthropometric fingerprint can reveal the user's identity. The system needs to be enhanced to handle the procedure to fulfill the user's requirements, such as quick processing time, nearly complete accuracy, no faults in the real procedure. Thus, in this chapter, we suggest the application of machine learning techniques to improve fingerprint classification procedures founded on the distinctiveness feature. Utilizing computer vision procedures during image pre-processing increases the quality of input images. As a result, feature extraction is extremely efficient and classification procedure is fast and precise.

The biometric systems are utilized for different tasks, such as computer login and building access. Fingerprint recognition is one of the most dominant techniques for personal recognition in all biometric structures (Phasuk et al., 2021). All over the world, fingerprint identification is accepted by a huge portion of the population due to its secure, fast and easy method of personal recognition. The fingerprint is the oldest human recognition and the most interesting which is utilized for individual recognition. In the initial 20th century, the fingerprint was officially accepted as a legal sign of recognition by law enforcement agencies.

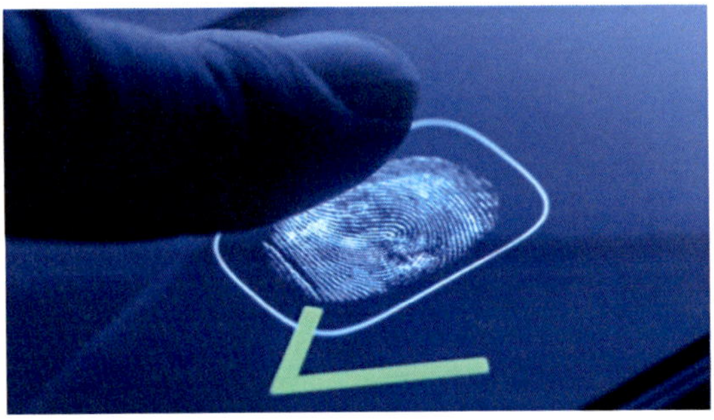

Figure 6.1. Fingerprint Scanner.

Source: *https://computer.howstuffworks.com/fingerprint-scanner.htm.*

On the base of this the system for automatic fingerprint identification for authentication and recognition. For instance, generally, there are two kinds of fingerprint recognition system: Automatic Fingerprint Authentication System (AFAS) and Automatic Fingerprint Identification/Verification System (AFIS); formed by developers and scientists lately (Wenxuan et al., 2019; Abiodun et al., 2019).

Acknowledgment of distinctive finger impression context relies on two important features like as:

- Individuality: The finger impression is distinctive of every person.
- Persistence: The major features of distinctive finger impressions don't alter with time.

The key parameters describing a digital fingerprint image are area, resolution, depth, geometric precision, number of pixels, etc.

Usually, fingerprints comprise valley and ridge patterns on the human finger's tips. Ridges (also termed ridge lines) are dark however valleys are bright (see Figure 6.2). Valleys and Ridges frequently run in parallel; occasionally they split and occasionally they terminate (Zwaan et al., 2010). Thanks to their continuity and distinctiveness, the usage of fingerprints is deliberated to be one of the most consistent techniques for personal verification.

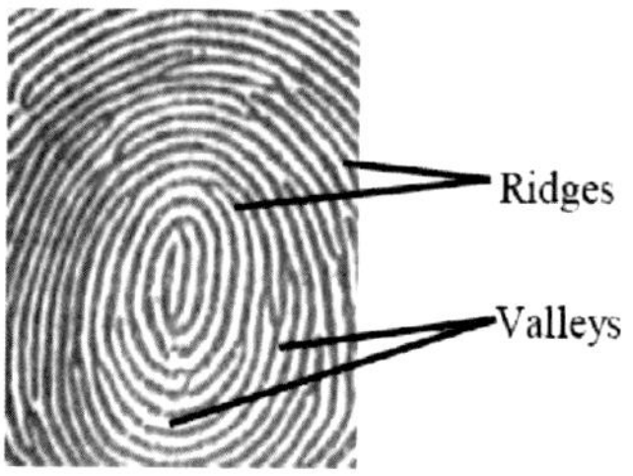

Figure 6.2. Ridges and valleys on the fingerprint image.

Source: *https://www.researchgate.net/figure/Figure2-Ridges-and-valleys-on-a-fingerprint-image-B-Two-Stage-Enhancement-Scheme-for_fig2_320868391.*

When examined at the worldwide level, the fingerprint pattern shows one or further regions where the ridgelines adopt distinctive shapes (considered through high curvature, regular termination, etc.). These regions (termed singularities or singular regions) might be categorized into three typologies: delta, loop, and whorl (see Figure 6.3) (Kotsiantis et al., 2007).

Fingerprint identification is widespread due to its various features like easily available several sources (10 fingers accessible for collection), ease in data collection. The usage of fingerprints had a record from the 19th century. It is all through this time Sir Francis Galton defined the fingerprint's features points with which it could be recognized and these were denoted as Galton points (Boykov & Jolly, 2001). The Galton Points provide the foundation on which a fingerprint identification system is formed. The Galton point's subset are utilized to signify the fingerprint image and these are mentioned as minutiae. Minutia states to several methods that the ridges could be discontinuous. For instance, a ridge could unexpectedly terminate (comes to an end) or could split into two ridges (bifurcation).

Mostly the fingerprint identification and classification procedures need a feature extraction phase for classifying salient features. The features mined from fingerprint images regularly had a direct physical complement (e.g., minutiae and singularities), however occasionally they are not directly linked to any physical characters (e.g., filter responses or local orientation image). Features might be utilized either for corresponding or their computation might assist as an intermediate phase for the other features derivation. For instance, certain preprocessing and improvement steps are regularly performed to shorten the minutiae extraction task (Tomin et al., 2106).

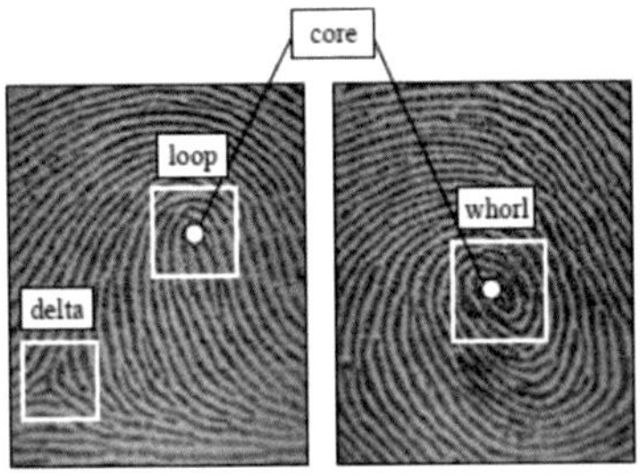

Figure 6.3. Core points and singular regions in fingerprint images.

Source: *https://www.researchgate.net/figure/a-Ridges-and-valleys-on-a-fingerprint-image-b-Singular-regions-white-boxes-and_fig1_220181660.*

6.2. APPLICATION OF FINGERPRINT IDENTIFICATION

During the crime scene, fingerprints play a significant part to recognize the criminals involved. CSI (crime scene images) are images obtained from the crime site. When a crime has happened, the investigator can gather both patent and latent fingerprints samples left behind by the perpetrator. The patent fingerprints have seemed through the naked eye, so they are merely photographed. However latent fingerprints are imperceptible and these samples are tougher to be visible (Phasuk et al., 2021). These samples could be raised by different methods. The usage of cyanoacrylate vapors attaches to prints and makes them observable in the normal light existence. This technique is much more challenging, thus usually in a crime scene, the investigators spread a fine dusting powder (black granular or aluminum dust) to the surface so fingerprints could be mined (Wenxuan et al., 2019). The dust attaches to the fingerprint after they use clear tape to take the fingerprints. Once the fingerprints are taken, they are scanned and kept in the digital image type. The fingerprints got from the crime site are accidentally made and these images are incomplete prints or noisy and hard to recognize.

6.3. RELATED WORK

Alpaydin (2018) showed a whole crime scene fingerprint identification structure utilizing deep machine learning with CNN (Convolutional Neural Network). Images are attained from crime scenes utilizing techniques extending from precision photography to complicated chemical and physical processing methods and kept as the database. Images gathered from the crime scene are generally partial and therefore difficult to classify. Appropriate enhancement techniques are needed for pre-processing the images of a fingerprint. Minutiae are mined from the images of a fingerprint. The characteristics of preprocessed records are fed into the CNN. The experimental outcomes confirmed on a database utilizing Open CV-Python displays high precision of 80% recognition of complete or partial fingerprints in the criminal record.

Wenxuan et al. (2019) suggested an algorithm that could efficiently enhance the location precision of the endpoints in the site area. By including neighborhood gathering into positioning reference, the technique enhances the positioning exactness of end positioning points, increases the whole indoor positioning outcome, and increases the positioning precision to a certain range. The simulation environment is formed to confirm the suggested algorithm. The experimental consequences show that the enhanced algorithm increases the positioning exactness of edge finding points to a certain range, and enhances the whole positioning exactness of end locating points.

Abiodun et al. (2019) provide readers with a vibrant understanding of the existing, and novel trend in ANN models that efficiently describe PR tasks to permit research attention and topics. Likewise, the complete review discloses the diverse areas of the achievement of ANN models and their usage to PR. In assessing the performance of ANN models performance, some statistical points for assessing the ANN model performance in several studies were accepted. Like as the usage of mean absolute error (MAE), the variance of absolute percentage error (VAPE), mean absolute percentage error (MAPE), root mean squared error (RMSE), and mean absolute error (MAE). The outcome displays that the existing ANN models like DBN, RBFN, CNN, RNN, SLP, MLP, SAE, MLNN, GAN, RBM, PNN, Transformer models, and Reservoir computing, are executing exceptionally in their application to PR chores. Thus, the study endorses the research emphasis on existing models and the new model's development simultaneously for more achievements in the field.

Serafim et al. (2019) give a method of splitting up the region of interest founded on CNN without pre-processing stages. The new method was assessed in two diverse architectures from state of the art, giving similarity indexes Distance of Hausdorff (5.92), Jaccard Similarity (96.77%), Dice coefficient (97.28%) superior to the classic techniques. The error rate (3.22%) was superior to five segmentation methods from the state of the art and presented improved results than additional deep learning methods, showing promising outcomes to recognize the region of interest with probable for application in systems founded on biometric identification.

Whang et al. (2014) suggested a method founded on an adaptive median filter for enhanced processing of fingerprint images and instinct removal of noise in the paper. The usage of adaptive median filtering to eliminate the instinct noise of the fingerprint image mostly comprises three steps. First, the adaptive median filter window size is prepared, and it's percent whether the middle pixel of the filter window in the image of a fingerprint is instinct noise. Second, the filter window size is determined founded on the maximum value, median value, and minimum value in the filter window. Lastly, median filtering is done on the fingerprint image in the filter window size got in the earlier steps, and the value of filter output is utilized in place of the window middle pixel value. The technique is verified on rolled fingerprint images adulterated by fingerprint-images and impulse noise and adulterated through impulse noise from a crime site.

6.4. EXISTING TECHNOLOGIES FOR FINGERPRINT ANALYSIS

A fingerprint is featured by ample and strong textural data. The textural properties on the surface of a live fingertip are reliable upon pore distribution, perspiration phenomenon, and skin elasticity. As a consequence, the pixels around and alongside the ridges of a live fingerprint show random and extensive variations in values of gray-level. The physical and material features of spoof fingers are consistent. The spoof and live fingerprints vary in inter-ridge distances, gray-level distribution, and ridge thickness (Senior et al., 2001). Thus, texture features-based techniques which could capture these alterations from fingerprint image properties are predicted to perform superior.

Spatial area pixel intensity value alterations generate texture arrays in an image. One of the features of texturing of image analysis utilizing texture properties. The present texture features founded fingerprint liveness detection techniques are grouped into the following categories:

i. Local texture features;
ii. Global texture features;
iii. Hybrid (global and local) texture features.

Machine learning is progressively showing its significance in parts of practical application (Voropai et al., 2018). Thus, scientists are researching and forming machine learning techniques that are becoming increasingly ideal. The one application which is attracting development is fingerprint classification. The method to the fingerprint classification issue is examined in Reference (Maio & Maltoni, 1996). The fingerprint is segmented into areas, this work would decrease the alteration of the element directions. A relation graph is made founded on the splitting up of the directional image. A perfect graph is utilized to likened it with the achieved graph, which could be modified to graph matching methods. Nyongesa et al. (2014) provide a fingerprint classification structure and its act in an identification scheme utilized a classifier of fuzzy neural network. The classification system is founded on fingerprint feature mining, which includes encoding the particular points together with their respective directions and positions got from an image of binaries fingerprint. Image examination is done in four steps, namely segmentation, directional image approximation, singular-point mining, and feature encoding. A technique is suggested by Zhang & Yan (2004) founded on singularities and ridges linking singular points. Due to images of low quality, it is very tough to get exact positions of particular features. The authors used curves features and ridge tracing analysis to categorize fingerprints. A machine learning procedure that takes a robust method for the fingerprints classification is SVM. This technique had produced a highly precise classification structure. The algorithm advantage is presented in the fingerprints classification into several classes. Li et al. (2008) utilized a combination of the naive Bayes technique with the SVM algorithm for fingerprints classification founded on the number of delta and core points on fingerprints. Several researchers had tried to extract particular points in the ridges' flow. Nyongesa et al. (2004) suggested a heuristic algorithm with distinctiveness for fingerprints classification; the drawback of these studies is not concentrating on enhancing the quality of the image. As they utilize features of distinctiveness points position. This takes to a great impact on the precision of the system. The Galton-Henry classification system is utilized to categorize fingerprints. This technique is presented by Nguyen (2019). They used rotation invariant distance, then relate this distance amongst the FingerCode trial set with the novel fingerprint outline. Throughout the classification procedure, the extraction of rotation-invariant distance took place in parallel with the training procedure. This provides the advantage of quick system time. Random forest (RF) is utilized to handle huge amounts of records and multi-class issues in classification (Everingham et al., 2010). This algorithm is a collective of randomized decision trees that correct the classification task. The most obvious application of RF is the discovery of human body parts from complex records (Nguyen et al., 2018). This application shows the feasibility of RF for machine learning issues in the actual world.

There are several known methods to increase the fingerprint classification structures performance. In this chapter, we suggested a mixture of traditional machine learning techniques and computer vision algorithms to treat such issues, which had effectively research in other areas (Derin & Elliott, 1987). Furthermore, we also designed and executed image processing techniques with denoising and raising edge utilizing the Canny technique.

6.5. PROPOSED FINGERPRINT IMAGE PRE-PROCESSING METHOD

6.5.1. Fingerprint Image Pre-Processing and Enhancement

There had been several earlier studies on resolving a noisy or partial image dataset. However, image pre-processing is a tough field although computer vision methods, such as image denoising, appear to be very promising.

Usually, image-denoising techniques are utilized to treat noisy images (Xiang et al., 2021). Denoising overwhelms the perturbations and improves the edges. This operation could generally be signified as image blurring, followed through the improvement of the edges. Thus, the subsequent image focuses on edges and repressed details, thus overwhelming the noise in the image.

A dataset is directly gathered through much noise, particularly noise with a tremendously high ratio of impulses, that brings a significant trial for image denoising. Thus, in the noise filter phase, we suggest a filter method founded on the CNN (convolutional neural network) (Haddad & Akansu, 1991). This procedure comprises two key steps. Initially, we develop a pre-processing stage for noisy images utilizing non-local information. Afterward, the pre-processed images are separated into patches and utilized for CNN training, taking to a CNN denoising model for forthcoming noisy images, to perceive the image noisy pixels and then flat them utilizing a Gaussian filter technique. In the CNN training phase, the pre-processed images are distributed into overlying patches. We utilize these patches as input for CNN. Our network had three layers; in every layer, we describe a set of operators and filters to produce mappings. The convolutional outcome of every patch is conforming to an n-dimensional feature map. We describe a convolutional layer to proposed increasing patches and redo them as a consequence image in the third layer. In this effort, the Gaussian algorithm (Canny, 1986) is utilized to filter noise with mask 3 × 3, after a canny algorithm is executed to increase the information edge. So, when the morphology operation had treated entire images, the outcome would be better. Image pre-processing stages are as follows (Breiman, 2001).

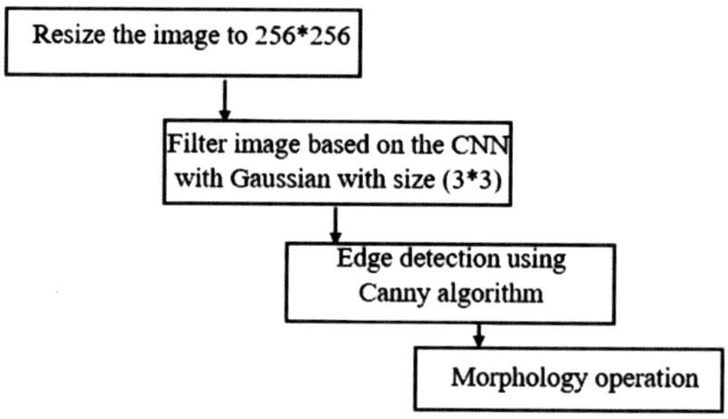

Figure 6.4. Block diagram of fingerprint image pre-processing.

Source: *https://www.researchgate.net/figure/Block-diagram-of-fingerprint-image-pre-processing-The-result-of-image-pre-processing-of_fig1_337187291.*

The outcome of image pre-processing of a fingerprint is revealed in Figure 6.5. Figure 6.5a,b display the consequences of the equalization histogram and noise filter founded on CNN, Figure 6.5c,d display the outcomes of edge detection and morphology.

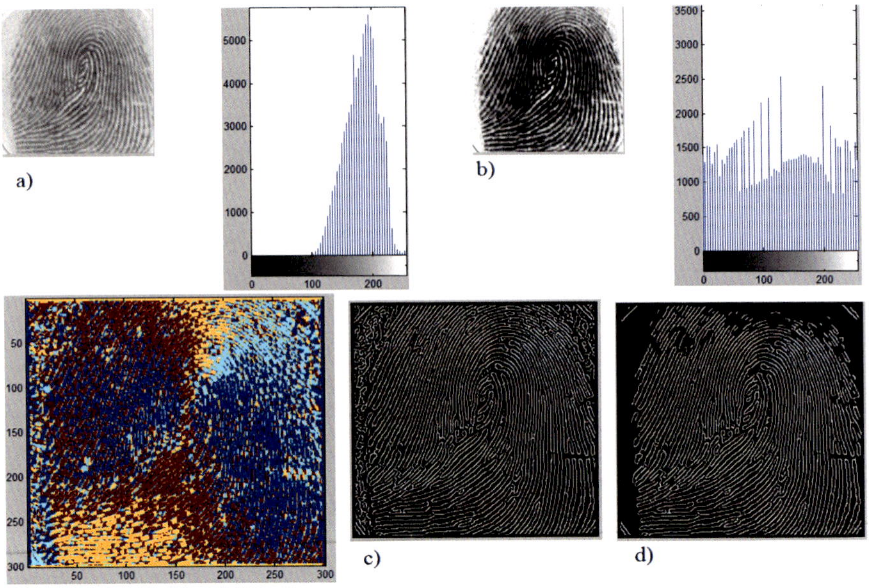

Figure 6.5. The outcome of image pre-processing of a fingerprint. (a,b) show the consequences of the equalization histogram and noise filter founded on CNN, (c,d) show the outcomes of edge detection and morphology.

Source: *https://www.researchgate.net/figure/The-result-of-image-pre-processing-of-a-fingerprint-a-b-show-the-results-of-the_fig2_337187291.*

6.5.2. Types of Fingerprint and Features Extraction

In this chapter, we categorize types of the fingerprint into three classes arch, loop, and whorl.

6.5.2.1. Arch

These are found in about 5% of the met fingerprints. The recognizing features of this arch are the overlapping shapes of fingerprints that create layers and had a mountain-like peak. Fingerprints arch are separated into numerous categories AE (a mixture of whorl and arch group, the distance from the midpoint to the intersection of eagles is below than 5 veins), AU, AR (the mixture of the loop group with arch, the distance from the midpoint to the intersection is below than 5 fringe lines), AS (the lines are loaded on top of each other, no intersection, unconcerned,) as presented in Figure 6.6 (Tarar et al., 2012).

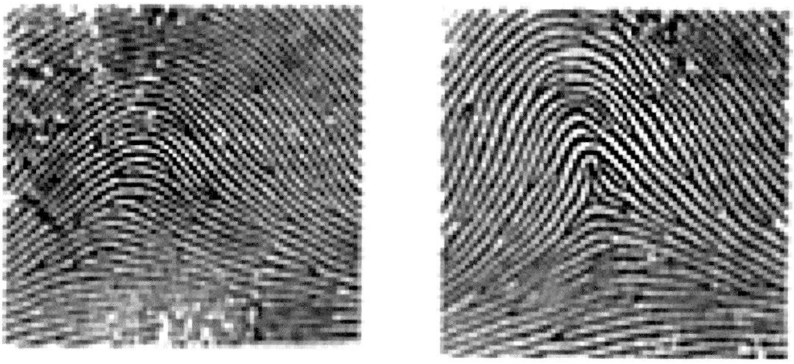

Figure 6.6. Example kinds of a fingerprint is an arch.

Source: *https://www.crime-scene-investigator.net/fbiscienceoffingerprints.html.*

6.5.2.2. Loop

It is termed the loop (could be seen in 60–65% of fingerprints globally) fingerprint due to its shape like a water wave having the following features the ridges make a regressive turn in loops, triangular in the midpoint, and an intersection. Separated into two kinds: *RL Radial Loop*: Upper of the triangle fronting the pinky finger. It seems like a stream of water moving downwards (on the little finger). This kind accounts for around 6% of fingerprints globally. *UL Ulnar Loop:* The upper of the triangle looks like the thumb. It is formed like a stream of water moving backward (thumb direction). This type only accounts for 2% of fingerprints globally. A loop pattern had only a single delta as shown in Figure 6.7. (Phasuk et al., 2021)

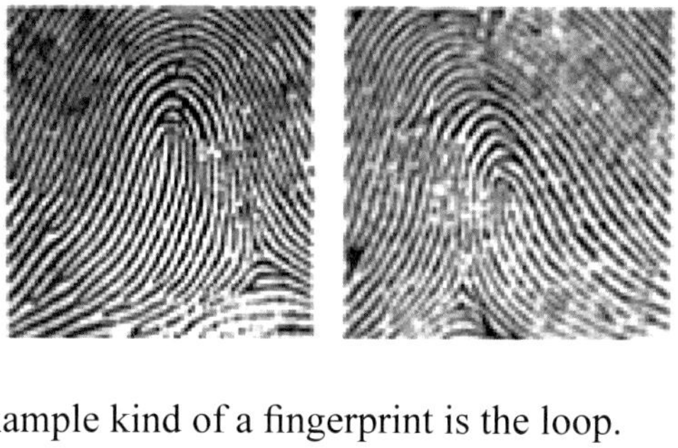

Figure 6.7. An example kind of a fingerprint is the loop.

Source: *https://www.researchgate.net/figure/Major-Fingerprint-Types-Whorl-Arc-Tent-Right-loop-Left-loop-and-Double-Loop_fig1_289246081.*

6.5.2.3. Whorl

This fingerprint merely accounts for around 25% to 35% of fingerprints globally. Whorl pattern recognition is that they have 1 circuit and 2 Delta (intersection) as exposed in Figure 6.8.

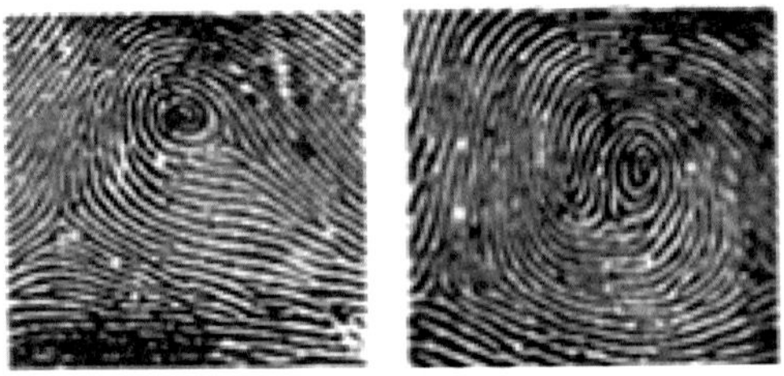

Figure 6.8. Example kinds of a fingerprint are whorl.

Source: https://sites.psu.edu/jlipton/2014/06/03/fingerprints-unique-to-us-all/.

6.5.2.4. Feature Extraction of Singularity

There are areas on fingerprint with rare structures related to other areas. They frequently had a parallel structure termed a singularity. There are two kinds of singularity core and delta. To excerpt singularity characteristics, we continue as follows:

- Stage 1. Input image then resize the image to 256 × 256.
- Stage 2. Fingerprint pre-processing and improvement.
- Stage 3. At every pixel, the gradient is computed in two directions x and y are G_x and G_y founded on the Formula (Phasuk et al., 2021):

$$\varphi = \frac{1}{2}\tan^{-1}\left[\frac{\sum_{i=1}^{W}\sum_{j=1}^{W} 2G_x(i,j)\,G_y(i,j)}{\sum_{i=1}^{W}\sum_{j=1}^{W}\left(G_x^2(i,j) - G_y^2(i,j)\right)}\right] \tag{1}$$

- Stage 4. Recognize singularity points utilizing the Pointcare index. Pointcare index at the pixel with organizes (i, j) is the total of the deviations of the direction of adjacent points, computed as follows in Equation (Wenxuan et al., 2019):

$$PC(i,j) = \sum_{k=0}^{N_p-1} \Delta(k) \tag{2}$$

$$\Delta(k) = \begin{cases} d(k);|d(k)| < \frac{\pi}{2} \\ d(k) + \pi; d(k) \leq -\frac{\pi}{2} \\ d(k) - \pi \end{cases} \quad d(k) = \varphi(x_{k-1}, y_{k+1}) - \varphi(x_k, y_k), \tag{3}$$

where ϕ is the incline at pixels in two directions. Founded on the Pointcare index, we could recognize singularity points as follows in Equation (Serafim et al., 2019):

$$PC(i,j) = \begin{cases} 0^0; (i,j) \text{ is not singularity} \\ 360^0; (i,j) \text{ is whorl} \\ 180^0; (i,j) \text{ is loop} \\ -180^0; (i,j) \text{ is delta} \end{cases} \quad (4)$$

- Stage 5. Save and generate fingerprint features vector.

6.6. CLASSIFICATION FINGERPRINT FOUNDED ON RANDOM FOREST AND DECISION TREE WITH SINGULARITY FEATURES

Fingerprint categorization is a multi-class classification issue. Labels are nominated as the novel input records for training a multi-classifier through the method of a supervised technique. In this work, there are three labels (arch, loop, and whorl). For the classification of fingerprints, a comparatively little number of features are mined from fingerprint images (Kim, 2003). Here, we select the positioning field base on the incline and recognize singularity features utilizing the Pointcare index as our arrangement. The machine learning procedures selected for the training module comprise Random Forest and Support vector machine (Kim, 2003; Yildiz et al., 2017) (Figure 6.9).

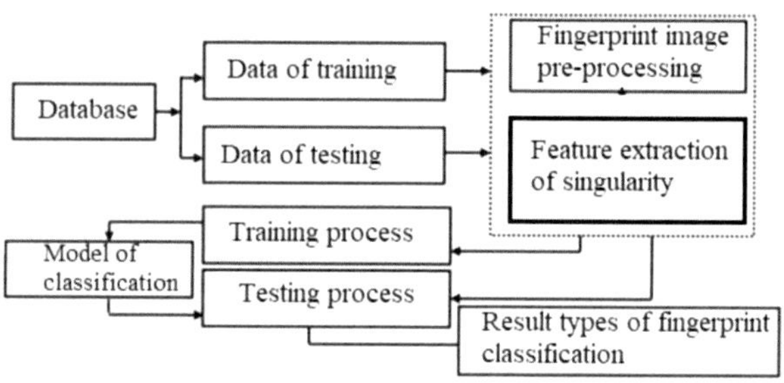

Figure 6.9. Block diagram for classification of fingerprint utilizing technique proposed.

Source: *https://www.researchgate.net/figure/Block-diagram-of-fingerprint-clas sification-using-method-proposed_fig6_337187291.*

REFERENCES

1. Abiodun, O. I., Jantan, A., Omolara, A. E., Dada, K. V., Umar, A. M., Linus, O. U., ... & Kiru, M. U. (2019). Comprehensive review of artificial neural network applications to pattern recognition. *IEEE Access*, *7*, 158820–158846.
2. Alpaydin, E. (2018). Classifying multimodal data. In *The Handbook of Multimodal-Multisensor Interfaces: Signal Processing, Architectures, and Detection of Emotion and Cognition, 2*, 49–69.
3. Boykov, Y. Y., & Jolly, M. P. (2001, July). Interactive graph cuts for optimal boundary & region segmentation of objects in ND images. In *Proceedings Eighth IEEE International Conference on Computer Vision, 1*, 105–112.
4. Breiman, L. (2001). Random forests. *Machine Learning*, *45*(1), 5–32.
5. Canny, J. (1986). A computational approach to edge detection. *IEEE Transactions on Pattern Analysis and Machine Intelligence*, *6*, 679–698.
6. Couturier, S., Erkman, L., Valera, S., Rungger, D., Bertrand, S., Boulter, J., ... & Bertrand, D. (1990). Alpha 5, alpha 3, and non-alpha 3. Three clustered avian genes encoding neuronal nicotinic acetylcholine receptor-related subunits. *Journal of Biological Chemistry*, *265*(29), 17560–17567.
7. Cristianini, N., & Scholkopf, B. (2002). Support vector machines and kernel methods: the new generation of learning machines. *Ai Magazine*, *23*(3), 31–31.
8. Derin, H., & Elliott, H. (1987). Modeling and segmentation of noisy and textured images using Gibbs random fields. *IEEE Transactions on Pattern Analysis and Machine Intelligence*, (1), 39–55.
9. Dijkstra, T., Milan, S., Vaclav, H., & Roger, B. (1994). Image processing, analysis and machine vision. *Neurocomputing*, *6*, 555.
10. Everingham, M., Van Gool, L., Williams, C. K., Winn, J., & Zisserman, A. (2010). The pascal visual object classes (voc) challenge. *International Journal of Computer Vision*, *88*(2), 303–338.
11. Fouad, S., Randell, D., Galton, A., Mehanna, H., & Landini, G. (2017). Epithelium and stroma identification in histopathological images using unsupervised and semi-supervised superpixel-based segmentation. *Journal of Imaging*, *3*(4), 61.

12. Gao, Q. (2013). An Experimental Study on the Accuracy Issue of Automatic User Verification Based on a Single Fingerprint Image. *International Journal of Scientific & Engineering Research*, *4*(10), 4–10.
13. Haddad, R. A., & Akansu, A. N. (1991). A class of fast Gaussian binomial filters for speech and image processing. *IEEE Transactions on Signal Processing*, *39*(3), 723–727.
14. Hu, H. H., Kuo, T. B. J., Wong, W. J., Luk, Y. O., Chern, C. M., Hsu, L. C., & Sheng, W. Y. (1999). Transfer function analysis of cerebral hemodynamics in patients with carotid stenosis. *Journal of Cerebral Blood Flow & Metabolism*, *19*(4), 460–465.
15. Kamat, A. A., Feng, S., Bogatcheva, N. V., Truong, A., Bishop, C. E., & Agoulnik, A. I. (2004). Genetic targeting of relaxin and insulin-like factor three receptors in mice. *Endocrinology*, *145*(10), 4712–4720.
16. Kim, K. J. (2003). Financial time series forecasting using support vector machines. *Neurocomputing*, *55*(1–2), 307–319.
17. Kotsiantis, S. B., Zaharakis, I., & Pintelas, P. (2007). Supervised machine learning: A review of classification techniques. *Emerging Artificial Intelligence Applications in Computer Engineering*, *160*(1), 3–24.
18. Lepetit, V., Lagger, P., & Fua, P. (2005, June). Randomized trees for real-time keypoint recognition. In *2005 IEEE Computer Society Conference on Computer Vision and Pattern Recognition,* 2, 775–781.
19. Li, J., Yau, W. Y., & Wang, H. (2008). Combining singular points and orientation image information for fingerprint classification. *Pattern Recognition*, *41*(1), 353–366.
20. Lowe, D. G. (2004). Distinctive image features from scale-invariant keypoints. *International journal of computer vision*, *60*(2), 91–110.
21. Maio, D., & Maltoni, D. (1996, August). A structural approach to fingerprint classification. In *Proceedings of 13th International Conference on Pattern Recognition*, 3, 578–585.
22. Nagaty, K. A. (2001). Fingerprints classification using artificial neural networks: a combined structural and statistical approach. *Neural Networks*, *14*(9), 1293–1305.
23. Nguyen, H. T. (2019). Fingerprints classification through image analysis and machine learning method. *Algorithms*, *12*(11), 241.

24. Nguyen, H. T. (2019). ROC curve analysis for classification of road defects. *BRAIN. Broad Research in Artificial Intelligence and Neuroscience*, *10*(2), 65–73.
25. Nguyen, H. T., Nguyen, L. T., & Dreglea, A. I. (2018). Robust approach to detection of bubbles based on images analysis. *International Journal of Artificial Intelligence*, *16*(1), 167–177.
26. Nguyen, T. H., Nguyen, T. L., Sidorov, D. N., & Dreglea, A. I. (2018). Machine learning algorithms application to road defects classification. *Intelligent Decision Technologies*, *12*(1), 59–66.
27. Nyongesa, H. O., Al-Khayatt, S., Mohamed, S. M., & Mahmoud, M. (2004). Fast robust fingerprint feature extraction and classification. *Journal of Intelligent and Robotic Systems*, *40*(1), 103–112.
28. Ou, M., Wang, J., Wu, Y., & Yi, J. (2020, September). Research and implementation of the HD video real-time edge detection system based on FPGA. In *Journal of Physics: Conference Series*, 1646(1), 012144.
29. Phasuk, S., Pairojana, T., Suresh, P., Yang, C. H., Roytrakul, S., Huang, S. P., ... & Liu, I. Y. (2021). Enhanced contextual fear memory in peroxiredoxin 6 knockout mice is associated with hyperactivation of MAPK signaling pathway. *Molecular Brain*, *14*(1), 1–17.
30. Rezaei, Z., & Abaei, G. (2017). A Robust Fingerprint Recognition System Based on Hybrid DCT and DWT. In *2017 24th National and 2nd International Iranian Conference on Biomedical Engineering,4,* 30–333.
31. Schulter, S., Leistner, C., Roth, P. M., Bischof, H., & Van Gool, L. (2011). On-line Hough Forests. In *BMVC*, 3, 1–11
32. Senior, A. (2001). A combination fingerprint classifier. *IEEE Transactions on Pattern Analysis and Machine Intelligence*, *23*(10), 1165–1174.
33. Serafim, P. B. S., Medeiros, A. G., Rego, P. A., Maia, J. G. R., Trinta, F. A., Maia, M. E., ... & Neto, A. V. L. (2019, July). A method based on convolutional neural networks for fingerprint segmentation. In *2019 International Joint Conference on Neural Networks, 4,* 1–8.
34. Strobl, C., Malley, J., & Tutz, G. (2009). An introduction to recursive partitioning: rationale, application, and characteristics of classification and regression trees, bagging, and random forests. *Psychological Methods*, *14*(4), 323.

35. Tarar, S., Kumar, A., Kumar, E., & Noida, G. (2012). A fingerprint watermarking algorithm to enhance privacy of fingerprint data. *WSEAS Transactions on Information Science and Applications*, *9*(8), 2224–3402.
36. Tomin, N. V., Kurbatsky, V. G., Sidorov, D. N., & Zhukov, A. V. (2016). Machine learning techniques for power system security assessment. *IFAC-PapersOnLine*, *49*(27), 445–450.
37. Voropai, N. I., Tomin, N. V., Sidorov, D. N., Kurbatsky, V. G., Panasetsky, D. A., Zhukov, A. V., ... & Osak, A. B. (2018). A suite of intelligent tools for early detection and prevention of blackouts in power interconnections. *Automation and Remote Control*, *79*(10), 1741–1755.
38. Wang, R., Han, C., Wu, Y., & Guo, T. (2014). Fingerprint classification based on depth neural network. *arXiv preprint arXiv:1409.5188*.
39. Wenxuan, B., Zhihong, F., Min, P., & Pu, W. (2019, April). Research on Indoor Edge Location Based on Location Fingerprint Recognition. In *2019 11th International Conference on Measuring Technology and Mechatronics Automation, 42,* 302–306.
40. Xiang, S., Liang, Q., Sun, W., Zhang, D., & Wang, Y. (2021). GSNet: Group Sequential Learning for Image Recognition. *Cognitive Computation*, *13*(2), 538–551.
41. Yildiz, M., Yanikoğlu, B., Kholmatov, A., Kanak, A., Uludağ, U., & Erdoğan, H. (2017). Biometric layering with fingerprints: template security and privacy through multi-biometric template fusion. *The Computer Journal*, *60*(4), 573–587.
42. Zhang, Q., & Yan, H. (2004). Fingerprint classification based on extraction and analysis of singularities and pseudo ridges. *Pattern Recognition*, *37*(11), 2233–2243.
43. Zwaan, L., de Bruijne, M., Wagner, C., Thijs, A., Smits, M., van der Wal, G., & Timmermans, D. R. (2010). Patient record review of the incidence, consequences, and causes of diagnostic adverse events. *Archives of Internal Medicine*, *170*(12), 1015–1021.

Chapter 7

Artificial Intelligence in Biometrics

CONTENTS

7.1. INTRODUCTION

Security has become the major issue of every company in the digital era. Each company has recognized that data is a valuable asset. As a result, companies use complex security methods to protect their company data. Various industrial titans, on the other hand, have had big data thefts in past recent years. Millions of consumers' personal information have been compromised as a result of cyber-attacks. As a result, companies are continuously on the lookout for better security solutions to existing security methods. Biometrics like iris scans and fingerprints are being used to verify mobile phone owners and authenticate employees at work. Companies may use biometrics to grant access to data for private information. For multi-factor identification, biometrics may be utilized with standard keywords or PINs. Furthermore, AI adoption would aid in the development of data-driven security systems (Boult et al., 2014). As a result, combining AI with biometrics would result in the construction of adaptive security models. AI systems that, if correctly programmed, may reduce "human error" situations and help people to make faster decisions by using cognitive methods. It is seen to be a very successful approach since it makes it difficult for hackers to break in and assures user protection.

The meaning of Biometrics is "the measurement of life." Anyone who uses biometrics can be identified by physiological methods (face picture, hand geometry, fingerprints, iris, and retinal) or behavioral characteristics that can be measured (keystroke rhythms, audio, signature). These are just a handful of the various techniques that are used in today's society to communicate. Practically speaking, many biometric systems operate based on 2nd phase procedures (Burt et al., 2019). A current user's data is recorded for the 1st time during Enrollment, which is the initial stage in creating the database. Data about a consumer is collected and passed via a method that converts the original data into a design that would be stored in the database alongside the individual's data. We should realize that the design, not the real data, is saved in the database, and the design includes relatively little information in the form of numbers. The 2nd stage is to recognize what you've done. When information is inserted for identification, a similar process is used to transform it into a design. This design is compared to the database, and if there is a +ve or –ve match, it is stated. The essential distinction between authentication (verifying an individual's identity) and recognition (identifying an individual's identity) should be understood. Authentication is a one-to-one validation, whereas recognition is a one-to-many validation (Manby et al., 2021).

Biometrics has a wide range of uses, including:

- **Biometric Passport:** Reduces passport piracy and forging; certain passports have RFIDs, that aid in the authentication of an applicant.
- **Military:** Each military around the globe has an information and data security division, and biometrics are used in this sector to assist secure secrets, detect fakers, and allow quick access to data when the company needs it.
- **Money Transactions:** Fingerprint sensors are now standard throughout all new ATMs, allowing for cardless transactions.
- **Airport security:** Ben Gurion International Airport in Israel, one of the busiest airports around the globe, installed biometric kiosks that allowed for efficient operation with no latency and reduced human mistakes, amongst many other things (Corrigan, 2017).

Biometrics is used in several aspects of daily life, yet there are several situations where it lacks the necessary information, including (Kalera et al., 2004):

- Artificial intelligence (AI) can fool conventional biometrics by utilizing artificial fingerprints. With artificial fingerprints, brute strength assaults may be planned, testing every possibility until none remain.
- One of the most important reasons is security; biometrics is safe but not infallible. Databases may be manipulated and may be fooled into giving the impression just like biometric is input because it matches with databases.
- Face detection may be defeated by twins.

Given that Artificial intelligence may outperform traditional biometric devices, what better approach to improve biometric equipment than to include the very danger and making it the finest arrow in the bow? When Artificial intelligence and biometrics are combined, safety and reliability may be multiplied tenfold (Corrigan, 2017).

7.2. HISTORICAL BACKGROUND

Boukhris et.al. (2011): this project looks at how the face-detection may be used to create virtual characters. Fraud prevention in multiplayer online video games is a problem, so this assists to mitigate the hazard to some level. Wavelet transformation and support vector machine (SVM) were among the approaches utilized by the researchers. SVM acts as a classifier and its effectiveness is greatly enhanced by wavelet transformation's characteristics utilized in photo processing. SVM is used to recognize objects. SVM works in a following way (Lanitis et al., 2009): utilizing a universal method, the data it receives is reorganized into a legible format and comprehensible to the program. The system's productivity is assessed using two failure rates: False Rejection Rate (FRR) and False Acceptance Rate (FAR). The data from the experiment shows that the average FAR and FRR are 5.5 and 2%, respectively. When compared to traditional biometric tests, this is a huge achievement. Because avatars have a minimal amount of facial gestures, this approach was restricted to 10 photos per input; however, when given to people, the precision and data gathering should be more quantifiable.

Key Interval Time (KIT) biometrics is analyzed using artificial intelligence and neural networks. The conventional keyboard is split into ten-word groups, which are supposed to represent the various keys that the 10 fingers would press. The User Rights and Integrity Enforcement Logic platform (URIEL) was created to evaluate Key Interval Time using neural networks. The User Rights and Integrity Enforcement Logic platform was permitted to reject or decide based on the credibility of the response during the initial anonymous human testing. In 82% of situations, the User Rights and Integrity Enforcement Logic platform might make a decision, although not in the remainder. As a result, the User Rights and Integrity Enforcement Logic platform will rather not remark than make a mistake, demonstrating the consistency and certainty of the answer's accuracy (Mazouni & Rahmoun, 2011). When every input was entered into the User Rights and Integrity Enforcement Logic platform, the precision of the result improved. When there are many submissions, the User Rights and Integrity Enforcement Logic platform utilizes the KOH (King of the Hill) technique, in which the strongest positive input has proclaimed the champion; if there are no positive inputs, the User Rights and Integrity Enforcement Logic platform unable to make a judgment. According to the data gathered, the User Rights and Integrity Enforcement Logic platform would correctly identify over 82% of all supplied inputs and would not provide a false positive rate of more than 2%, which is a highly promising outcome in the area of biometrics. The User Rights and Integrity Enforcement Logic platform rejects to make a choice that may be modified in subsequent adjustment in lower than eighteen percent of the inputs. The most important thing to remember is that Artificial intelligence has a lot of potential in psychological biometrics.

Online signature identification is much more interactive than offline signature identification since it depicts the route taken by the subject when signing. This may assist to reduce forging to a certain level, although expert forgers may still get around it (Kalera et.al. 2004). The signature stroke is difficult to establish an offline signature identification. Signature is a behavioral feature that is acquired over a long time via consistent practice; it's not a psychological biometric. During the investigation, two databases A & B were assumed: "A" is a complete offline database wherein signatures were scanned in 256 grey shades and published in PNG format, and "B" is a digitally signed database. Because database B was digitized, the pen coordinates had to be converted into X-Y coordinates firstly, which required a lengthy time. The signatures were run through a GSC feature system, which included Structural, Concavity, and Gradient. It transforms the signatures to binary format and uses a similarity metric to equate the 2. The 2/3 of each writer's signatures were utilized for training, with the remaining 8 being utilized for assessment. The findings for databases A & B are plotted on a graph among FAR and FRR. Database A had an Equal Error Rate (ERR) of 21.90%, whereas Database B had an ERR of 32.37%. The accuracy of a completely offline setup was 93.18%, however, combining Dynamic Plane Wrapping (DPW) with this technique may enhance the performance of the methodology by a factor of 10 (Lee & Park, 2021).

Aging has a significant impact on a person's appearance for understandable reasons; it is natural and permanent. Throughout this discipline, researchers are attempting to describe face biometrics in this manner that the identification isn't affected by aging. Individuals age in various ways; certain exhibit signs of aging early on, while others may not exhibit signs of aging until later in life (Lanitis et al., 2009). The argument is that there is no such thing as a universal aging trend that pertains to all people. We divide the participants into two groups: A and B, & C and D, with an average age gap of 1.6 years for A and B & fifteen years for C and D. When coping with varied age divisions, a 12% reduction in productivity is observed, according to the findings. In addition, the upper facial region is more susceptible to refusal because it ages than the lower facial region. Data-driven study in such a field is useless because we already know that the consequences of aging manifest themselves differently in various people. Only sophisticated setups capable of changing face templates while considering the factors of aging and ensuring that the template is compatible with the person's existing visage may avoid this. This is time-consuming and appears to be unattainable because external variables influence the aging procedure and exacerbate the effects. For this reason, using a cutting-edge Artificial intelligence system for face identification is quite useful. The requirement of the hour is for a system that considers the impact of external circumstances while also defining face representations that contain discriminating and also time-invariant characteristics.

Cooperative Biometrics Abnormality Detection system (C-BAD). C-BAD may be used to efficiently monitor an individual's actions, with any irregularities identified automatically being communicated to security staff. Companies are quite well ready for outside threats, but inner trust issues are something for which they are unprepared. Employers are unable to adequately image their workers as threats, which makes them vulnerable to hostile harassment. A calm workplace atmosphere can't be formed if bosses see their staff as possible threats (Kocher et al., 2006). In a study of hundred information security setups, 45% of them were hostile, while 10% were criminal. One out of every two of such individuals was an information technology expert, with 19% being top-level system administrators and 31% being assistant system admins. Insiders may only be captured if the proof is skewed; otherwise, neither of the colleagues will testify to one of their own committing a crime. Functioning before/after planned work time, creating duplicates or publishing, or accessing data that is not allowed to them are examples of these anomalies. Because monitoring requires providing C-BAD complete permission to the files, and sensitive material can't be made available to anyone else, confidential files of an enterprise are not monitored in this way (Pascu et al., 2015). The main access of C-BAD is restricted to the agency's leader, and the main system server is equipped with artificial intelligence or a rule-dependent system that determines whether a disorder is worth investigating or not, removing the human element and ensuring complete integrity and the publication of an impartial result free of any manipulation.

A lot of people, a limited specimen size, and high dimensionality are three features of a standard Biometric Authentication (BA). The single specimen biometrics identification issue, where only oe specimen is supplied and should be utilized for future verifications, is a key challenge in Biometric Authentication (Yao et.al., 2007). This usually results in a low identification rate. A superior technique depending upon characteristic level biometrics fusion is given here, in which two forms of biometrics, namely the facial characteristic (a non-touch biometric) and the palm feature (a touch biometric), are combined (A usual contact biometric). Gabor-dependent image pre-processing and Principal Component Analysis (PCA) techniques are used to extract the discriminant characteristics. Such that, an individual's distinguishing characteristics on such two fronts are segregated and saved as identification data in the database (Liang et al., 2020). Then, to produce feature-level fusion, a distance-based dissimilarity grading technique is devised. The use of a big database for facial and palm identification enhances recognition rates considerably, which is to be anticipated because the larger the database the program is given, the more distinguishing traits or distinctions it may integrate into the recognition database. This assists to overcome the limitations of single-specimen biometrics and improve the BA system's efficiency. It also suggests that facial and palm print biometrics has a significant relationship.

Geometry detection techniques in biometrics recognize a person based on the geometry of a sub-region of the body. Hand geometry authentication is because every individual has a distinct hand form that doesn't even vary after a certain time (Boult et.al. 2014). Because the form of the hand simply may not be enough to distinguish persons, other characteristics must be defined as well. To successfully differentiate two hands to the maximum degree of accuracy, photographs of each hand should be acquired from a similar range and at an identical angle concerning the hand. This has to be addressed throughout the enrollment procedure. Although it is recognized that comparison occurs on a one-to-one basis during recognition, the subject should be compared to every database item during recognition The planar projective invariant characteristics acquired from the hand biometric are combined with standard biometrics gained from face detection in this software. The essential point here is that both of such traits may be recorded in a single photo that simplifies storage and eliminates data waste during enrollment. This not only allows for the storage of a huge database of persons but also ensures that identification is not hampered in any way (Manby, 2021).

Massive technology use has provided us with several benefits; however one of the drawbacks is balancing availability and security. The outcomes of keystroke dynamics are Authentication (Am I who I say I am?) and Recognition (Who am I?). The effectiveness of this method is increased when it is combined with keyword or password security (Wong et al., 2001). Not only does the intruder need to remember the passcode, as well as the pace at which the consumer types it in. That biometric kind is highly cost-effective without sacrificing security. For the time interval system, a Real-Time Clock (RTC) or Clock/Counter Time Chip (CTC) might be used. This program written in Visual C++ utilizes the K-Nearest Neighboring (KNN) rule to organize information; unidentified input information is allocated to the next closest neighbor in the sequence; this may appear to minimize system integrity, however at the resolution when every information set is saved, KNN would have no impact on system protection, but it would minimize identification time. Another method that may be used is the Multi-Layer Perceptron for Artificial Neural Networks (ANNMLP). The average approved acceptance and false acceptance rates for KNN and ANN-MLP, accordingly, are 84.63% and 1.03% for KNN and 99% and 29% for ANN-MLP. It is entirely up to the consumer to decide which approach to utilize.

This article addresses fusion approaches, which include combining more than one biometric information set for a person to improve identification. Genetic Algorithm (GA), Particle Swarm Optimization (PSO), Brute Force Search (BFS), Support Vector Machines (SVM), and Adaptive Neuro-Fuzzy Systems (ANFIS) are the five decision level techniques that are commonly utilized (Mazouni & Rahmoun, 2011). The results are derived utilizing Unconstrained Cohort Normalization (UCN) without and with applying normalization, and the efficiency is measured utilizing Equal Error Rate (EER). Every such approach excels in gathering and evaluating information, and each works differently in different limitations and circumstances. Relating one object to another may seem irrational, yet under some circumstances; the comparisons might make complete sense to people who understand information. The following are the findings: Genetic Algorithm and Particle Swarm Optimization beat other approaches because of their ability to scan a big database. Owing to the use of neural networks to obtain the appropriate variables for the optimal Flexible Input System, ANFIS's findings are similar to those of Genetic Algorithm & Particle Swarm Optimization. Since the real and impostor distributions contain so many intersections, Support Vector Machines have the lowest efficiency (Purgason & Hibler, 2012).

Authorship authentication by using text classification is done. In the traditional technique, a writer's material is scanned, and 200 of the most frequently utilized terms are selected as content words and physically deleted, while the remaining words are considered and evaluated. However, Artificial intelligence treats every response as a 130-dimensional vector. The relative frequency is represented by every element in this vector. Furthermore, to discriminate among distinct collections, a version of an exponential gradient is utilized (Koppel et.al. 2011). This is a unique type of biometric that isn't commonly utilized. However, this approach is used by a lot of copyright software. To the untrained eye, this may appear to be a useless type of biometric, yet many fans of literature all around the globe would agree that copyright is no less of a crime for which one may be prosecuted. Artificial Intelligence simplifies the "unmasking" process, which involves isolating the strongest discriminators and differentiating the task of determining the handwriting of both, Standard methods can't successfully out-think a human brain's inventiveness, however an electronic brain, if not outwit, may at least find similarities among two works (Sudhir et al., 2019).

AI plays a significant part in the everyday lives of the increasing population, as well as its expanded accessibility profits. The banking sector is currently one of India's most important sectors. Banks across the world are undergoing rapid digital transformations to get more benefits, such as enhanced synergies, cost-effectiveness, business performance, and reducing cyber risks, among other things, and the latter is one of the top goals (Sudhir et al., 2019). Because of the large amounts of money and personal data held by banks, it's become a major target for hackers. Banks are constantly threatened by financial losses, regulations, and brand damage, as seen by regular headlines about data breaches. Furthermore, corruption has compelled bank employees and customers to expand their security systems. Without a question, banks and other financial institutions must be cyber-secure from the inside out. Tangible fingerprints have been supplanted by modern biometrics and further improved in the years after the advent of AI.

Huawei is the largest supplier to NEC (serving 14 countries) for 50 countries, more than any company. Artificial intelligence systems are employed by just over half of "established democracies," compared to only 37% of authoritarian states, according to the research. However, the technologies are not always misused, according to the study (Burt et al., 2019).

The technology might eventually permit espionage agencies to recognize persons who utilize cameras put on far-off roofs and disused planes, as per Spy Advanced Research Projects Agency, the CIA's research arm. Face detection and other forms of biometric technology have advanced significantly in recent years, but even the most unique approach is less trustworthy and lacks of clear understanding of their subject (Corrigan, 2017). Even if the person stands close by and looks right into the camera, face recognition systems might make errors. The intelligence community, on the other hand, aims to eliminate such limitations in two ways: by gathering more information and designing algorithms that recognize people using multiple data kinds.

AI technology is rapidly gaining traction throughout the world. Starting with deep visuals that bridge the gap between fact and fiction, advances continue to be made, and advanced methods may compete with the best multi-player poker players on the planet (Feldstein et al., 2012). Organizations utilize artificial intelligence to improve analytical processing, and city governments utilize artificial intelligence to monitor traffic congestion and smart energy measurement. Nonetheless, a growing number of countries are using advanced monitoring tools to watch, monitor, and monitor individuals to obtain a range of policy objectives, certain of which are legitimate, certain of which are violating human rights, and many of which are somewhere in the center. To effectively address the effects of new technology, we must first understand where such tools are utilized and how they've been utilized. Regrettably, such information is scarce. To further clarify things, this chapter includes an AIGS (Artificial Intelligence Global Monitoring Index), which is the first of its type in research. The index gathers factual data on Artificial Intelligence surveillance utilize in 176 countries throughout the world. It makes no distinction among legal and criminal applications of Artificial Intelligence surveillance. Rather, the goal of the research is to show how new capacities to observe and trace individuals or systems alter government capability (Wong et al., 2001).

7.3. FINDINGS

The results may be summarized as a comparison of several techniques, which combine artificial Intelligence with biometrics. In the area of biometrics, the effectiveness of a technology is measured using two variables: False Acceptance Rate (FAR) and False Rejection Rate (FRR). If the biometric machine grants accessibility to anyone who is not authorized to access the data, this instance is considered an FAR. On the other hand, FRR refers to a system that denies entry to an authorized user. Every one of the studies mentioned above shows that biometric systems function better with artificial intelligence (AI). Whereas the actual dilemma is which technique to use when combining biometrics with AI, a few of the fusion strategies discussed in Literary Survey Number 9 deal with integrating multiple kinds of data into a singular one so that neural networks may operate successfully on it. Because of the approach used to combine the data about one person, Genetic Algorithm and Particle Swarm Optimization outperforms the others. As a result, we may deduce that many approaches may be utilized, but the careful study is required to identify which is most appropriate for the activity in question (Purgason & Hibler, 2012).

Tangible signatures are superseded by current biometrics and further refined in the years since the advent of AI. The behavioral anomaly may now be detected and security concerns may be avoided with greater precision. Robots and cybercriminals may now be identified more easily by this tech. Lengthy Biometric Recognition is being investigated by the intelligence agencies. AI tech is rapidly advancing around the world. The introduction of sophisticated movies that obscure the boundary between truth & fiction is leading us to advanced methods that are the best players in multiplayer poker on the planet (Liang et al., 2020).

Artificial intelligence and cognitive science are two distinct categories, each with its own set of tools and aims. AI is a branch of software engineering concerned with the creation and organization of clever experts as computer programs. The goal of AI is to understand the underlying principles of smart behavior that equally apply to real and artificial systems. The bulk of the research is mathematical or statistical, and a large portion of the literature is procedural.

Psychological science is a very multidisciplinary area that benefits from Artificial Intelligence, but also reasoning, brain research, and other sociological and organic scientific subspecialties. Cognitive science's overarching goal is to grasp and demonstrate human knowledge by employing the complete range of findings and systems of the fundamental categories. A broad range of methodologies from the technical, behavioral, sociological, and biological sciences are used, as one might anticipate. There are active research groups in both Artificial Intelligence and psychology, but they would generally provide different types of reports for journals and conferences in both fields (Wangsuk & Anusas-Amornkul, 2013).

7.4. SUMMARY

From unlocking our devices to registering our presence, biometrics are regular aspects of several of our lifestyles. Even the Indian government uses biometrics entered during Aadhaar registration to identify people. Most of this suggests that biometrics is now in usage; the challenge now is to upgrade the current tech to make such devices function more efficiently. AI may help make things go quicker and much safer. All of these fields of Artificial intelligence, such as machine learning, data analytics, and neural networks play a critical role in developing biometric tech. Biometric devices nowadays are incapable of withstanding brute force attacks. After reading several research articles, I concluded that while a lot of hard work is being put into evaluating different methods for enhancing the stability of biometric machines, the outcomes are not achieving the scientific community at a stage where they may have a major effect, and if they perform, technological giants are cautious of incorporating Artificial intelligence into their products. With several benefits like Artificial Intelligence has, there are only so many drawbacks; much research has been conducted and is now being done and I am certain would be done tomorrow to overcome these obstacles. The next most difficult step will be to explain such tech to a layperson that is a simpler task (Chang, 2012).

Large corporations were harmed by security breaches that harmed e-mail addresses, personal information, and passcodes. According to cybersecurity experts on numerous occasions' passcodes are very vulnerable to attacks on loan card data, private information compromise, and social security numbers. All of the above is how biometric logins are advantageous to cybersecurity.

AI, in particular, has made our lifestyle easier. Although if we didn't realize it at the time, Alexa's general applications have made their way into our homes' tech. Here's a brief breakdown of how Artificial intelligence would affect us at every stage in life (Traore et al., 2014).

Constant maintenance of equipment is a huge expense for manufacturers, thus switching from responsive to predictive maintenance has become a requirement. Through employing artificial neural networks and advanced Artificial Intelligence techniques to detect property failure and early alert to engineers, Artificial intelligence has been able to save businesses money and time (Deutschmann et al., 2013).

REFERENCES

1. Abuhamad, M., Abusnaina, A., Nyang, D., & Mohaisen, D. (2020). Sensor-based continuous authentication of smartphones' users using behavioral biometrics: A contemporary survey. *IEEE Internet of Things Journal*, *8*(1), 65–84.
2. Al-Dori, A. S. M. (2021). Touchscreen-based Smartphone Continuous Authentication System (SCAS) using Deep Neural Network. *Turkish Journal of Computer and Mathematics Education (TURCOMAT)*, *12*(11), 2382–2391.
3. Arel, I., Rose, D. C., & Karnowski, T. P. (2010). Deep machine learning-a new frontier in artificial intelligence research [research frontier]. *IEEE Computational Intelligence Magazine*, *5*(4), 13–18.
4. Ayeswarya, S., & Norman, J. (2019). A survey on different continuous authentication systems. *International Journal of Biometrics*, *11*(1), 67–99.
5. Balyen, L., & Peto, T. (2019). Promising artificial intelligence-machine learning-deep learning algorithms in ophthalmology. *The Asia-Pacific Journal of Ophthalmology*, *8*(3), 264–272.
6. Bleecker, E. R., Busse, W. W., FitzGerald, J. M., Ferguson, G. T., Barker, P., Sproule, S., ... & Tanaka, A. (2019). Long-term safety and efficacy of benralizumab in patients with severe, uncontrolled asthma: 1-year results from the BORA phase three extension trial. *The Lancet Respiratory Medicine*, *7*(1), 46–59.
7. Boakes, M., Guest, R., Deravi, F., & Corsetti, B. (2019). Exploring mobile biometric performance through identification of core factors and relationships. *IEEE Transactions on Biometrics, Behavior, and Identity Science*, *1*(4), 278–291.
8. Boukhris, M., Mohamed, A. A., D'Souza, D., Beck, M., Amara, N. E. B., & Yampolskiy, R. V. (2011, July). Artificial human face recognition via Daubechies wavelet transform and SVM. In *2011 16th International Conference on Computer Games, 3,* 18–25.
9. Boult, J. K. R., Cummings, C., Waterton, J. C., Ulloa, J., ... & Robinson, S. P. (2014). Tumour biomechanical response to the vascular disrupting agent ZD6126 in vivo assessed by magnetic resonance elastography. *British Journal of Cancer*, *110*(7), 1727–1732.
10. Burt, C., Edwards, S. M., Buntjer, J. B., Jackson, R., Bentley, A. R., Lage, J., Byrne, E., ... & Hickey, J. M. (2019). The effects of training

population design on genomic prediction accuracy in wheat. *Theoretical and Applied Genetics*, *132*(7), 1943–1952.

11. Chang, T. Y. (2012). Dynamically generate a long-lived private key based on password keystroke features and neural network. *Information Sciences*, *211*, 36–47.
12. Chen, W., Shahabi, H., Shirzadi, A., Hong, H., Akgun, A., Tian, Y., ... & Li, S. (2019). Novel hybrid artificial intelligence approach of bivariate statistical-methods-based kernel logistic regression classifier for landslide susceptibility modeling. *Bulletin of Engineering Geology and the Environment*, *78*(6), 4397–4419.
13. Corrigan, J. (2017). Project Maven Uses Machine Learning to Go Through Drone Video Feeds, but That's Just the Beginning, Air Force Lt. Gen Shanahan Said. *Nextgov, November, 2, 321.*
14. Das, S., Dey, A., Pal, A., & Roy, N. (2015). Applications of artificial intelligence in machine learning: review and prospect. *International Journal of Computer Applications*, *115*(9), 4–10.
15. Deutschmann, I., Nordström, P., & Nilsson, L. (2013). Continuous authentication using behavioral biometrics. *IT Professional*, *15*(4), 12–15.
16. Dimiduk, D. M., Holm, E. A., & Niezgoda, S. R. (2018). Perspectives on the impact of machine learning, deep learning, and artificial intelligence on materials, processes, and structures engineering. *Integrating Materials and Manufacturing Innovation*, *7*(3), 157–172.
17. Feldstein, S. B., Garfinkel, C. I., Waugh, D. W., Yoo, C., & Lee, S. (2012). Observed connection between stratospheric sudden warmings and the Madden□Julian Oscillation. *Geophysical Research Letters*, *39*(18), 40–10.
18. Frank, M., Biedert, R., Ma, E., Martinovic, I., & Song, D. (2012). Touchalytics: On the applicability of touchscreen input as a behavioral biometric for continuous authentication. *IEEE Transactions on Information Forensics and Security*, *8*(1), 136–148.
19. Gangurde, H., Gulecha, V., Borkar, V., Mahajan, M., Khandare, R., & Mundada, A. (2011). Swine influenza A (H1N1 virus): a pandemic disease. *Systematic Reviews in Pharmacy*, *2*(2), 110.
20. Ghahramani, Z. (2015). Probabilistic machine learning and artificial intelligence. *Nature*, *521*(7553), 452–459.

21. Gonzalez Viejo, C., Torrico, D. D., Dunshea, F. R., & Fuentes, S. (2019). Emerging technologies based on artificial intelligence to assess the quality and consumer preference of beverages. *Beverages*, *5*(4), 62.
22. Helm, J. M., Swiergosz, A. M., Haeberle, H. S., Karnuta, J. M., Schaffer, J. L., Krebs, V. E., ... & Ramkumar, P. N. (2020). Machine learning and artificial intelligence: definitions, applications, and future directions. *Current Reviews in Musculoskeletal Medicine*, *13*(1), 69–76.
23. Hoang, N. D., & Pham, A. D. (2016). Hybrid artificial intelligence approach based on metaheuristic and machine learning for slope stability assessment: A multinational data analysis. *Expert Systems with Applications*, *46*, 60–68.
24. IYER, A. P., Karthikeyan, J., KHAN, M. R. H., & Binu, P. M. (2020). An analysis of artificial intelligence in biometrics-the next level of security. *J Crit Rev*, *7*(1), 571–576.
25. Kalera, M. K., Srihari, S., & Xu, A. (2004). Offline signature verification and identification using distance statistics. *International Journal of Pattern Recognition and Artificial Intelligence*, *18*(07), 1339–1360.
26. Kocher, M. S., Tucker, R., Ganley, T. J., & Flynn, J. M. (2006). Management of osteochondritis dissecans of the knee: current concepts review. *The American Journal of Sports Medicine*, *34*(7), 1181–1191.
27. Koppel, M., Schler, J., & Argamon, S. (2011). Authorship attribution in the wild. *Language Resources and Evaluation*, *45*(1), 83–94.
28. Lanitis, S., Behranwala, K. A., Al-Mufti, R., & Hadjiminas, D. (2009). Axillary metastatic disease as presentation of occult or contralateral breast cancer. *The Breast*, *18*(4), 225–227.
29. Lee, Y., & Park, J. (2021). Using Big Data to Prevent Crime: Legitimacy Matters. *Asian Journal of Criminology*, *21*, 1–20.
30. Liang, Y., Samtani, S., Guo, B., & Yu, Z. (2020). Behavioral biometrics for continuous authentication in the internet-of-things era: An artificial intelligence perspective. *IEEE Internet of Things Journal*, *7*(9), 9128–9143.
31. Lu, L., & Liu, Y. (2015). Safeguard: User reauthentication on smartphones via behavioral biometrics. *IEEE Transactions on Computational Social Systems*, *2*(3), 53–64.
32. Manby, B. (2021). The Sustainable Development Goals and 'legal identity for all: First, do no harm.' *World Development*, *139*, 105343.

33. Mazouni, R., & Rahmoun, A. (2011). On comparing verification performances of multimodal biometrics fusion techniques. *International Journal of Computer Applications*, *33*(7), 24–29.
34. Nichols, J. A., Chan, H. W. H., & Baker, M. A. (2019). Machine learning: applications of artificial intelligence to imaging and diagnosis. *Biophysical Reviews*, *11*(1), 111–118.
35. Panch, T., Szolovits, P., & Atun, R. (2018). Artificial intelligence, machine learning and health systems. *Journal of Global Health*, *8*(2), 1–20.
36. Pascu, L. F., Nastase, G., & Pascu, B (2015). Potential problems of global scientific research, technological development and innovation. *The USV Annals of Economics and Public Administration*, *15*(3), 19–26.
37. Peng, G., Zhou, G., Nguyen, D. T., Qi, X., Yang, Q., & Wang, S. (2016). Continuous authentication with touch behavioral biometrics and voice on wearable glasses. *IEEE Transactions on Human-Machine Systems*, *47*(3), 404–416.
38. Purgason, B., & Hibler, D. (2012). Security through behavioral biometrics and artificial intelligence. *Procedia Computer Science*, *12*, 398–403.
39. Rashidi, H. H., Tran, N. K., Betts, E. V., Howell, L. P., & Green, R. (2019). Artificial intelligence and machine learning in pathology: the present landscape of supervised methods. *Academic Pathology*, *6*, 1-17.
40. Samtani, S., Kantarcioglu, M., & Chen, H. (2021). A multi-disciplinary perspective for conducting artificial intelligence-enabled privacy analytics: Connecting data, algorithms, and systems. *ACM Transactions on Management Information Systems (TMIS)*, *12*(1), 1–18.
41. Schclar, A., Rokach, L., Abramson, A., & Elovici, Y. (2012). User authentication based on representative users. *IEEE Transactions on Systems, Man, and Cybernetics, Part C (Applications and Reviews)*, *42*(6), 1669–1678.
42. Shah, P., Kendall, F., Khozin, S., Goosen, R., Hu, J., Laramie, J., ... & Schork, N. (2019). Artificial intelligence and machine learning in clinical development: a translational perspective. *NPJ Digital Medicine*, *2*(1), 1–5.

43. Sudhir, P., Dinesh, P. A., Suma, S. P., & Srinivasa, G. (2019). Characteristic study of combined effects of MHD and coriolis force on free convection in a rectangular cavity with isotropic and anisotropic porous media. *Mapana Journal of Sciences*, *18*(3), 13–31.
44. Traore, I., Woungang, I., Obaidat, M. S., Nakkabi, Y., & Lai, I. (2014). Online risk-based authentication using behavioral biometrics. *Multimedia Tools and Applications*, *71*(2), 575–605.
45. Viejo, C. G., Fuentes, S., Howell, K., Torrico, D., & Dunshea, F. R. (2018). Robotics and computer vision techniques combined with non-invasive consumer biometrics to assess quality traits from beer foamability using machine learning: A potential for artificial intelligence applications. *Food Control*, *92*, 72–79.
46. Vijai, C., & Wisetsri, W. (2021). Rise of Artificial Intelligence in Healthcare Startups in India. *Advances In Management*, *14*(1), 48–52.
47. Wangsuk, K., & Anusas-amornkul, T. (2013). Trajectory mining for keystroke dynamics authentication. *Procedia Computer Science*, *24*, 175–183.
48. Willis, A. D. (2019). Rarefaction, alpha diversity, and statistics. *Frontiers in Microbiology*, *10*, 2407.
49. Wong, F. W. M. H., Supian, A. S. M., Ismail, A. F., Kin, L. W., & Soon, O. C. (2001, November). Enhanced user authentication through typing biometrics with artificial neural networks and k-nearest neighbor algorithm. In *Conference Record of Thirty-Fifth Asilomar Conference on Signals, Systems and Computers*, 2, 911–915.
50. Yampolskiy, R. V., & Govindaraju, V. (2008, April). Behavioral biometrics for verification and recognition of malicious software agents. In *Sensors, and Command, Control, Communications, and Intelligence (C3I) Technologies for Homeland Security and Homeland Defense, 6943*, 694303.
51. Yang, Y., Guo, B., Wang, Z., Li, M., Yu, Z., & Zhou, X. (2019). BehaveSense: Continuous authentication for security-sensitive mobile apps using behavioral biometrics. *Ad Hoc Networks*, *84*, 9–18.
52. Yao, Y. F., Jing, X. Y., & Wong, H. S. (2007). Face and palmprint feature level fusion for single sample biometrics recognition. *Neurocomputing*, *70*(7–9), 1582–1586.

Chapter 8

Countering the Presentation Attacks in Recognition of Face, Iris, and Fingerprint

CONTENTS

8.1. INTRODUCTION

Biometric authentication is a method of automatically identifying people depending on their behavior, chemical characteristics, and physical characteristics. This innovation has emerged as a significant mechanism for connection regulation in numerous new applications, where traditional methodologies, such as knowledge-based (i.e., keywords) or tokens (i.e., smart cards), may be inefficient due to their ease of sharing, loss, theft, or manipulation (Jia et al., 2013). Biometric data are rapidly being utilized as the primary authentication feature for accessibility management, as well as in conjunction with conventional authentication procedures as a "step-up authentication" component in 2-factor or 3-factor systems of authentication.

Fingerprint, face, and iris are the most often employed biometric features in this scenario. Indeed, the attribute to be employed is chosen based on factors such as ease of calculating biometric parameters, performance, universality, and complexity of manipulating the system (Jia et al., 2013). Yet, one drawback of these characteristics is that a forger may create a synthesized replica that could be provided to the biometric detector to bypass the authentication procedure. Presentation attack detection (PAD), spoofing detection, and liveness detection are terms used in the research to describe the procedures that safeguard the biometric system from this form of assault. We'll refer to this as "presentation attack detection" from now (Gottschlich, 2016).

Spoofing biometric identification is a notion that is considerably earlier than biometrics altogether. An impersonation effort mentioned in the Genesis Book, depending on the display of a goat's fur placed on Jacob's hand to simulate qualities of Esau's skin such that Jacob would've been honored by Isaac, can be found. The depiction of how to reproduce somebody's fingerprint utilizing a gelatin and wax mold offered by Austin Freeman in his mystery book "The Red Thumb Mark" is a fictional illustration that is remarkably genuine. It was published in 1907, as well as the technology detailed in it is still utilized to spoof fingerprint detectors nearly a century later. It's worth noting that this depiction surfaced only 4 years following Scotland Yard embraced biometrics and well before the first biometric sensor reached the market (Galbally, 2012).

Presentation attacks are always a concern in biometrics, according to current research studies and public challenges like LivDet (www.livdet.org). Facial identification systems, according to Boulkenafet et al. (2016) are subject to presentation attacks, having an equivalent error rate (associated with separating presentation attacks from actual samples) as higher as 9%. The problem persists in fingerprint-based identification systems, having an average categorization error rate of 2.9%. Iris-based identification, which is widely regarded as among the most trustworthy biometrics, requires a more efficient PAD approach. The average categorization error rate in recent initiatives in this region is still about 1%.

Just a few real incidents demonstrate the concern, in addition to laboratory verification of the biometric system's susceptibility to assault. A physician from the facility of mobile healthcare services and urgency has been captured red-handed by cops in the tiny town of Ferraz de Vasconcelos, on the eastern side of Sao Paulo, Brazil, in a fraud that used silicone fingers to override an authentication protocol and indicate the existence of many workmates (Gragnaniello et al., 2015).

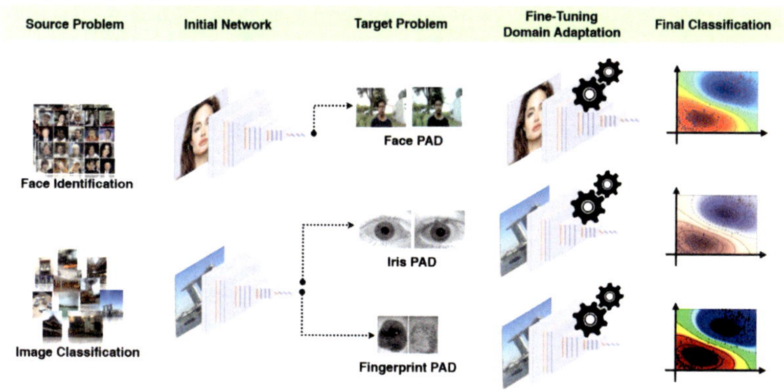

Figure 8.1. Fingerprint, iris, and face detection systems in a generalized flowchart.

Source: *https://www.taylorfrancis.com/chapters/edit/10.1201/b22524–11/counteracting-presentation-attacks-face-fingerprint-iris-recognition-allan-pinto-helio-pedrini-michael-krumdick-benedict-becker-adam-czajka-kevin-bowyer-anderson-rocha.*

In 2014, the Brazilian Federal Security probed an identical instance in which employees at Figure 8.1. Typical pipeline exploited in this activity. Initially, network topologies, which were initially presented for other challenges, are fine-tuned individually using suitable PAD samples from various datasets, resulting in discriminative characteristics. Finally, classifiers are taught to distinguish between genuine images of fingerprints, irises, faces, and their assault variants (Lee, 1996).

They were accused of employing silicone fingers to overcome a timed attended biometric technology at Paranagua Harbor inside the Brazilian state of Parana. The biometric hacking team of Chaos Computer Club successfully hacked the Touch ID of Apple's iPhone just days after its release in Germany, indicating that a biometric system lacking proper protection is unacceptable as a dependable accessibility management technique. Other incidents of people using 3-D masks to spoof surveillance systems to modify their perceived age or ethnicity can be reported (Kumpituck et al., 2017).

Once we glance at the literary works and assess the methodologies to avoid presentation attacks using the three modalities described above, we notice that even the most impressive in contexts of mistakes and execution endeavor or cost almost always share a fascinating function: those who refer to a family of methodologies known as data-driven characterization methodologies. Approaches relying on data characterization, so according to Pinto et al. (2015), merely use data from a baseline biometric sensor to seek for signs of artifacts in a previously obtained biometric template. Such techniques are preferred since they are simple to integrate with the current detection system, requiring no additional hardware or requiring human intervention to identify attempted assaults.

Even though current techniques based on this concept have resulted in high identification rates, we mention that certain aspects must still be considered when assessing a PAD methodology, such as various kinds of attacks, a diverse range of devices to carry out attempted assaults, and assaults aimed directly at various sensors. Another factor that is sometimes neglected is that so many detection systems are custom-tailored to certain sorts of presentation assaults, a process known as feature hand-crafting. With the rise of deep learning techniques and their effectiveness in activities such as picture classification, speech recognition, and language processing, we set out in this chapter to use data-driven approaches to utilize deep learning techniques for identifying presentation assaults (Rattani & Ross, 2014). In these instances, the biometric developer is in charge of selecting an acceptable design for training and PAD using just the data that is already accessible. We feel that this sort of approach is the next logical step in the development of powerful presentation assault sensors and that if correctly built, they could better handle the difficult cross-dataset issue. Whenever a system is prepared with data through one sensor and then evaluated on data from another sensor, the cross-dataset situation occurs. The main pipeline we use in this chapter is depicted in Figure 8.1. When creating the ultimate classifiers to differentiate between legitimate photos of faces, fingerprints, and irises from their stationary counterparts, we begin using pre-trained deeper neural networks and adjust them individually for every modality (fingerprint, face, and iris) with various datasets (Schwartz et al., 2011).

8.2. DETECTION OF FACE PRESENTATION ATTACK

Behavior modeling of user (i.e., small face movements, eye blinking), methodologies that necessitate extra hardware (e.g., depth sensors and infrared cameras), methodologies relying on usage collaboration (e.g., task questionnaire), and eventually data-driven characterization strategies, which would be the target of our task (Sequeira et al., 2014).

This section begins with a discussion of frequency-based techniques, which are techniques that analyze artifacts that are more evident in the domain of frequency. This concept was supported by early research but we currently have several studies that back up the usefulness of this method in identifying facial spoofing. A facial spoofing identification method, based on the assumption that photos' faces are small from actual ones, and their emotions and positions are unchanging. The authors developed a threshold-based judgment technique relying on the power rate of the higher frequencies in the 2-D Fourier spectrum to identify photo-based attempted assaults depending on these data. The approach described by Li et al. (2004) has a major drawback in that the higher frequency elements are influenced by light, making this frequency range excessively noisy (Li et al., 2004). Tan et al. (2010) used the variation in photo variability in the high-middle spectrum to decrease this impact. This is accomplished by the use of Gaussian (DoG) bandpass filtering Difference, which preserves as much information as feasible while avoiding the introduction of aliasing or noisy artifacts.

Pinto et al. (2015) proposed a method for overcoming the lighting effect while functioning in the frequency range. Rather than utilizing image pixel measurements directly, the researchers suggested a facial anti-spoofing approach for identifying video-based attempted assaults depending on Fourier's evaluation of the noise signature derived from movies. After separating the noise signal contained in the video clips, the researchers converted the data to the Fourier domain and utilized the visually rhythmic approach to collect the most relevant frequency parts to identify an attempted assault by combining temporal and spectral information. Similar authors built on this method in a much more current paper, employing the notion of visual codebooks to make use of the temporal, spatial, and spectral information from noise signature. The new approach, according to the scientists, allowed them to identify several sorts of assaults, including printed mask-based attempted assaults.

Lee et al. (2013) developed an anti-spoofing approach based on video imaging-based cardiac pulse measures. The authors added a threshold-based judgment level depending on the entropy measure to prior work presented by Poh et al. (2010). It was computed utilizing Individual Component Analysis from the energy spectrum derived from normalized RGB channels following removing cross-channel noise produced by environmental Independent Component Analysis (ICA).

Texture-based methods are yet an additional expressive area of facial anti-spoofing techniques identified in the literature. Such methods, in general, make use of textural cues introduced in false biometric data throughout presentation and creation to the biometric detector under assault (e.g., blurring, printing, and aliasing effects). Tan et al. (2010) developed a texture-based technique for detecting assaults with printed pictures depending on the surface roughness difference between a real face and an attempted attack. The writers use Sparse Low-Rank Bilinear Logistic Regression techniques to estimate the reflectance and brightness of the picture under investigation and categories it. Peixoto et al. (2011) built on their study by include measurements for varied lighting situations.

Kose et al. (2013) assessed a method relying on reflectance to identify assaults done using printed masks, identical to Tan et al. (2010). The Variational Retinex algorithm was used to split the pictures into reflectance and illumination components.

Micro textures were used for facial spoofing identification by Maatta et al. (2010), who were motivated by the analysis of printing artifacts and variations in reflecting light while comparing actual samples with presentation attacks. Relying on the Histogram of Oriented Gradients (HOG), Local Binary Pattern (LBP), and Gabor wavelets, the authors suggested a fusion technique. Likewise, Schwartz et al. (2011) presented an approach that uses multiple properties of the pictures to create a holistic picture of the face capable of revealing an attempted attack (e.g., shape, color, and texture of the face).

Maatta et al. (2010) looked at how several versions of the LBP operator may be employed. $\chi 2$ histogram comparisons, linear discriminant analysis (LDA), and support vector machine (SVM) were used to classify the histograms obtained from such descriptors.

The use of static masks to execute face spoofing attacks has also been discussed in the research. Erdogmus et al. (2013) used face information from four participants to investigate a database containing six kinds of assaults. The authors utilized two methods depending on Gabor-phase-based similarity as well as a Gabor wavelet measure to identify attempted assaults.

For identifying different forms of assaults, Pereira et al. (2013) developed a score-level fusion method. The authors implemented the Q statistics to analyze the interdependence among classifiers after training them with various datasets. Pereira et al. (2013) offered an anti-spoofing approach built on the dynamic pattern, which would be a spatiotemporal variant of the fundamental LBP in a follow-up paper.

Garcia et al. (2015) presented an anti-spoofing technique relying on the identification of Moire patterns, which arise when digital grids overlap. The scientists utilized a peak-detector method built on maximum thresholding to discover these patterns, with high peaks indicating an attempted assault. Patel et al. (2015) developed a presentation attack identification approach built on pattern detection of Moire (also known as M-LBP), which employs the LBP descriptor multi-scale version.

Tronci et al. (2011) combined two types of algorithms, video-based and static analysis, to leverage the motion information and hints retrieved from the scene. The static analysis includes many visual properties such as Gabor textures, color, and edge, while the video-based analysis integrates basic measures that are motion-related like variation in facial expression, eye blink, and lip movement. Using a stationary face identification system, Anjos et al. (2011) presented a method for identifying photo-based assaults. The strength of relative mobility among the background and face region, according to with authors, could be utilized as a hint to identify legitimate access from attempted assaults, since the motion fluctuations among the background and face regions have a higher correlation in attempted attacks.

Wen et al. (2015) developed an imaging distortions analysis (IDA)-based face spoof detection method that describes several characteristics like color variety, specular reflection, chromatic moment, and blurriness. These characteristics are combined to create feature vectors, that are being utilized to create an ensemble classifier, each of which is trained to identify a certain sort of attempted assault.

Kim et al. (2015) presented a technique for detecting attempted assaults depending on the single picture diffusion speed. Local diffusion speed patterns, particularly local speed patterns via Total Variation (TV) flow, are defined by the authors and utilized as feature vectors to train a linear classifier utilizing the SVM to assess if either a particular face is fake. In response, Boulkenafet et al. (2016) presented a color texture analysis-based anti-spoofing approach. Essentially, the authors extract characteristic descriptions from multiple color spaces to do a micro-texture analysis using color-texture information from the chrominance and luminance channels.

8.3. DETECTION OF FINGERPRINT PRESENTATION ATTACK

There are two kinds of PAD fingerprint techniques: software and hardware-based solutions. Techniques in the first category employ data from other sensors to collect artifacts that show a spoofing assault that occurs outside of the fingerprint photo. Software-based approaches rely exclusively on the data collected by the fingerprint recognition system's biometric sensor.

Galbally et al. (2012) developed a collection of attributes for fingerprint presentation assault identification depending on multiple quality metrics (i.e., ridge continuity, directionality, or ridge strength), that were fed into a classifier of Linear Discriminant Analysis (LDA).

Gragnaniello et al. (2013) developed an anti-spoofing approach relying on the Weber Local Descriptor (WLD), which works in conjunction with other texture descriptors like Local Binary Pattern Descriptor (LBP) and Local Phase Quantization (LPQ). Even whether examined alone or in combination with LBP, the research findings show that LPQ and WLD complement each other, but that their combined use can substantially increase their discriminating capacity.

Jia et al. (2013) developed a spoofing identification technique relying on Multi-Block Local Ternary Patterns (MBLTP), which was influenced by prior work centered on the LBP descriptor. The LTP descriptor is computed depending on averaging numbers of block subgroups instead of single pixels, about the researchers, that made it less vulnerable to noise because it is dependent on a 3-value code interpretation and mean value of block subgroups instead of single pixels.

Ghiani et al. (2013) suggested using Binarized Statistical Image Features (BSIF), a textured binary descriptor based on the LPQ and LBP techniques. Essentially, the BSIF descriptor trains a filter set from natural picture data, resulting in descriptors that are more suited to the situation. The LPQ descriptor was also used by similar researchers to discover a characteristic space that was unaffected by blurring impacts.

Gottschlich (2016) presented a different approach relying on a filter learning convolution comparative pattern. The researchers calculate the discrete cosine transform (DCT) using rotation unchanging patches and calculate their binary patterns by matching DCT coefficients pair to identify fingerprint faking. These patterns were compiled into a histogram and fed into a linear classifier that is SVM.

In the operational phase, Rattani et al. (2014) proposed a technique for automatically adapting a liveness sensor to novel spoofing materials. On an anti-spoofing technology, the suggested technique aims to mitigate the privacy risk caused by novel spoof materials. The scientists suggested a new material scanner that is specifically designed to identify novel spoof materials, emphasizing the necessity to train the system with the newly discovered material.

Rattani et al. (2014) developed an automated adaption anti-spoofing system relying on Weibull-calibrated SVM (W-SVM), which included a new material detection and open-set fingerprint spoofing detection. The open set fingerprint spoofing detection was trained using attributes dependent on anatomical, textural, and physiological characteristics, while the new material detector was constructed utilizing a multi-class W-SVM comprised of an ensemble of binary SVMs and pairs of 1-Class.

Gragnaniello et al. (2015) suggested a fingerprint spoofing identification method that used both frequency and spatial information to retrieve local image behavior and relevant amplitude contrast, which were synthesized by analyzing the phase of certain chosen transform coefficients produced by the short-time Fourier transform (STFT). The data was utilized to train a linear classifier SVM using a bi-dimensional contrast-phase histogram.

Kumpituck et al. (2017) used the LBP operator and a wavelet decomposition method-based anti-spoofing scheme. The researchers retrieve LBP histograms from multiple wavelet sub-band pictures in this study, which are then combined and fed into an SVM classifier. The scientists also looked at a more traditional technique, which involves estimating energy from wavelet sub-bands rather than LBP histograms. In addition to obtaining successful performance with state-of-the-art techniques, the wavelet LBP descriptor produced higher discrimination than LBP and wavelet-energy descriptors used individually.

Lastly, Nogueira et al. developed a fingerprint antispoofing approach relying on the idea of pre-trained convolutional neural networks, which differs from standard modeling in that it employs texture patterns to describe fingerprint pictures. The authors start by acquiring the network weights for fingerprint spoofing identification using the known architectures of CNN in the research, such as VGG and AlexNet.

8.4. DETECTION OF IRIS PRESENTATION ATTACK

The earliest work on iris spoofing detection dates from the 1990s when Daugman (1993) addressed the validity of certain iris recognition structure threats. In that paper, he suggested utilizing the Fast Fourier Transform to authenticate the higher frequency spatial intensity to identify such moves.

Options for iris liveness detection, as according to Czajka (2016), can be classified into four categories as a Cartesian outcome of two aspects: type of model of the object under dynamic or static test and type of passive or active measurement. Passive alternatives imply that the item is not triggered beyond what is required to obtain an iris picture for recognition. As a result, no additional hardware is usually needed to identify an attempted strike. Active solutions attempt to boost an eye and monitor its response. It usually necessitates the addition of some additional hardware components. As a result, the method can identify an attempted strike utilizing just one static picture from the biometric sensor, or it must use a series of pictures to examine identified dynamic characteristics. Hardly static and passive methods are discussed in this section, as they are the priority of this chapter.

Pacut et al. (2006) proposed three iris liveness detection methods depending on picture frequency spectrum assessment, regulated pupil dynamics, and light reflection from the cornea in their paper. These methods were tested using paper printouts from various printers and printout carriers, and they were found to be capable of fooling two industrial iris recognition structures. A narrow hole was created in the pupil's place, and this tactic was adequate to fool industrial iris recognition processes used in their research. The governed pupil dynamics and light reflections acquire zero for both False Rejection and Acceptance Rate, according to the experimental findings acquired on the assessment set of 77 pairs of live and fake iris pictures. Two commercial cameras, on the other hand, were unable to accurately recognize every iris paper printing.

Galbally et al. (2012) suggested a method depending on 22 picture quality indicators for instance; focus, pupil dilation, and occlusion. To choose the best characteristics, the researchers utilized sequential floating attribute selection, which was then fed into a quadratic discriminant classification.

The writers used the BioSec baseline to legitimize their approach, which includes print-based iris spoofing threats. Sequeira et al. (2014) used picture quality measures and three distinct classifying methods to validate their task on the BioSec and Clarkson benchmarks and introduce the MobBIOfake benchmark, which consists of 800 iris pictures. Sequeira et al. (2014) built on earlier work by adding a feature shortlisting step to get a better description for detecting an intended strike. The writers also used iris division to get the iris contour and then adapted the attribute extraction methods to the non-circular iris regions that resulted.

Wei et al. (2008) used three texture steps to solve the issue of iris-texton feature, iris edge sharpness (ES), and iris liveness detection for characterizing the graphic primitives of IT – iris texture, and using chosen characteristics relying on CM co-matrix. They wore color and contoured contact lenses with falsely iris in specific. The studies indicated that the ES characteristic produced outcomes that were analogous to state-of-the-art techniques at the time, and which the CM and IT measures surpassed the state-of-the-art methodologies.

To detect printable irises, Czajka (2016) suggested a frequency analysis-based alternative. Spikes in the frequency range were linked to frequent trends in printed specimens. This technique, which was adapted to obtain a near-zero false rejection rate, was capable of detecting 95% printable irises.

Iris spoofing detection has been also investigated using texture assessment. The winners used three texture identifiers in the MobILive iris spoofing detection competition: Binary Gabor Pattern (BGP), LPQ, and LBP. Sun et al. (2013) lately suggested a hierarchical visual codebook-based (HCV) structure for iris image classification. The HVC uses two established bag-of-words prototypes to encrypt the pattern primitives of iris pictures. The technique outperformed the competition in terms of iris recognition tasks and other iris spoofing detection.

Doyle et al. (2013) suggested a viable alternative relying on modified local binary patterns (mLBP) descriptor. The writers demonstrate that, while pattern data derived by the mLBP descriptor can be used to get good categorization outcomes, the technique's performance suffers substantially. When evaluated on two different databases, the authors obtain roughly 83–96% accuracy. When the descriptor was evaluated only on a single database, the accuracy drops to 42–53%. This cross-dataset verification is found in many validation configurations for detecting presentation attacks. Yadav et al. (2014) built on prior research by looking at how textured and soft contact lenses affect iris recognition.

Yadav et al. (2014) suggested an anti-spoofing technique relying on Eulerian Video Magnification (EVM) to boost the delicate phase data in the eye area. The researchers postulated a judgment rule relying on combined phase information that was implemented to the phase component utilizing a sliding window strategy for identifying the rate of phase transformation over time.

8.5. INTEGRATED FRAMEWORKS FOR DETECTING PRESENTATION ATTACKS

Galbally et al. (2012) suggested a particular strategy relying on 25 picture quality characteristics for synchronously detecting attack attempts in fingerprint, face, and iris biometric structures. When compared to state-of-the-art techniques devoted to single methods, assessments conducted on famous baselines for three methods demonstrate that the proposed method is extremely competitive.

Gragnaniello et al. (2013) used the SID (Shift-Invariant Descriptor), the BoVW (Bag-of-Visual-Word) model, DAISY (105), and SIFT (Scale-Invariant Feature Transform) to analyze numerous nearby descriptors for iris, fingerprint-, and face-based biometrics, as well as the inquiry of successful descriptors.

Menotti et al. (2015) demonstrated that combining filter and architecture optimization improves understanding of how such strategies interact for fingerprint, iris, and face PAD, as well as outperforming the best-recognized strategies on multiple standards.

8.6. DATASETS AND METRICS

The standards (datasets) and chosen accuracy estimators regarded in this research are described in this chapter. All of the sets of data utilized in this chapter were openly accessible to us, and we presume that the same is true for other scientists who contact their creators instantly. The most frequently utilized baselines for evaluating presentation attack tracking for fingerprints, iris, and face are the databases that make up our testing environment. We chose to obey these divides because all of the standards had already been partitioned by their innovators into testing and training subgroups. We split each training sub-group into two disjoint subgroups numerous times to execute cross-validation-based instruction to improve the winning model's generalization capabilities and reduce overfitting. The findings presented in the following sections are those acquired from testing sets. The following subsets describe all sets of data concisely (Ojala et al., 2002).

8.6.1. Benchmarks for Video-Based Face Spoofing

CASIA Face Anti-Spoofing and the Replay-Attack databases are used in this section to analyze the efficiency of PAD techniques for facial modality. This set of data includes five multiple forms of attempted threats using falsified specimens of various characteristics (Pacut & Czajka, 2006).

8.6.2. Replay-Attack

This baseline includes 50 multiple subjects' short video footage of both video-based targets and valid accesses. Every individual was documented with a regular web camera in two exercises, one in a regulated atmosphere and the other in an undesirable atmosphere, to create valid access video clips. After which, utilizing three approaches, spoofing efforts were created (Peixoto et al., 2011):

1. Mobile target: video clips on an iPhone display were shown to the acquisition sensor, and these video clips were also chosen to take with the iPhone;
2. High-definition target: employing the iPad display, high-resolution video clips and pictures were displayed to the acquisition sensor;
3. Print target: the acquisition sensor was introduced with hard prints of high-resolution digital images, which were published on a Triumph-Adler DCC 2520 color laser printer.

8.6.3. CASIA

This standard was created using specimens from 50 people. Authentic pictures were captured with three distinct sensors with varying acquisition quality: a long-time-used USB camera, a recently purchased USB camera, and a Sony NEX-5 camera. The picture resolution was either 1920×1080 pixels (Sony sensor) or 640×480 pixels (both webcams). The researchers trimmed Sony pictures to 1280720 pixels. Subjects were inquired to blink throughout the procurement. There were three types of presentation attacks conducted out (Phan et al., 2016):

1. Cut photo attack: eyes were cut out of paper printing, and an assailant hiding behind an artifact imitated blinking behavior whenever the Sony sensor was used to acquire the video.
2. Video attack: high-resolution genuine video clips were exhibited on an iPad display with a resolution of 1280×720 pixels.
3. Warped photo attack: high-resolution pictures were published on copper paper and video clips were captured using a Sony sensor; the printable pictures were twisted to emulate face micro-movements.

The information from 20 areas of study was used to create a training set, whereas the existing specimens (from 30 subjects) were used to create the testing set (Poh et al., 2010).

8.6.4. Benchmarks for Fingerprint Spoofing

This chapter used two sets of data from Liveness Detection Competitions (LivDet). LivDet is a sequence of worldwide competitions that use a standardized testing protocol and significant amounts of spoof and live specimens to document attack techniques for iris and fingerprint. All of the contests are accessible to all industrial and academic institutions that have bio-metric liveness detection software and systems (Rousson et al., 2003).

REFERENCES

1. Bharadwaj, S., Dhamecha, T. I., Vatsa, M., & Singh, R. (2013). Computationally efficient face spoofing detection with motion magnification. In *Proceedings of the IEEE Conference on Computer Vision and Pattern Recognition Workshops*, 12(1), 105–110.
2. Boulkenafet, Z., Komulainen, J., & Hadid, A. (2016). Face spoofing detection using color texture analysis. *IEEE Transactions on Information Forensics and Security*, *11*(8), 1818–1830.
3. Czajka, A. (2016). Iris liveness detection by modeling dynamic pupil features. In *Handbook of Iris Recognition*, 3, 439–467.
4. Daugman, J. G. (1993). High confidence visual recognition of persons by a test of statistical independence. *IEEE Transactions on Pattern Analysis and Machine Intelligence*, *15*(11), 1148–1161.
5. Erdogmus, N., & Marcel, S. (2014). Spoofing face recognition with 3D masks. *IEEE Transactions on Information Forensics and Security*, *9*(7), 1084–1097.
6. Ortega-Garcia, J., Fierrez, J., Alonso-Fernandez, F., Galbally, J., Freire, M. R., Gonzalez-Rodriguez, J.,, & Savran, A. (2009). The multiscenario multienvironment biosecure multimodal database (bmdb). *IEEE Transactions on Pattern Analysis and Machine Intelligence*, *32*(6), 1097–1111.
7. Galbally, J., Alonso-Fernandez, F., Fierrez, J., & Ortega-Garcia, J. (2012). A high performance fingerprint liveness detection method based on quality related features. *Future Generation Computer Systems*, *28*(1), 311–321.
8. Galbally, J., Ortiz-Lopez, J., Fierrez, J., & Ortega-Garcia, J. (2012, March). Iris liveness detection based on quality related features. In *2012 5th IAPR International Conference on Biometrics, 41,* 271–276.
9. Garcia, D. C., & de Queiroz, R. L. (2015). Face-spoofing 2D-detection based on moiré-pattern analysis. *IEEE Transactions on Information Forensics and Security*, *10*(4), 778–786.
10. Ghiani, L., Hadid, A., Marcialis, G. L., & Roli, F. (2013, September). Fingerprint liveness detection using binarized statistical image features. In *2013 IEEE Sixth International Conference on Biometrics: Theory, Applications and Systems, 543,* 1–6.
11. Gottschlich, C. (2016). Convolution comparison pattern: an efficient local image descriptor for fingerprint liveness detection. *PloS One*, *11*(2), e0148552.

12. Gragnaniello, D., Poggi, G., Sansone, C., & Verdoliva, L. (2013, September). Fingerprint liveness detection based on weber local image descriptor. In *2013 IEEE Workshop on Biometric Measurements and Systems for Security and Medical Applications*, *34*, 46–50.
13. Gragnaniello, D., Poggi, G., Sansone, C., & Verdoliva, L. (2015). An investigation of local descriptors for biometric spoofing detection. *IEEE Transactions on Information Forensics and Security*, *10*(4), 849–863.
14. Jia, X., Yang, X., Zang, Y., Zhang, N., Dai, R., Tian, J., & Zhao, J. (2013, June). Multi-scale block local ternary patterns for fingerprints vitality detection. In *2013 International Conference on Biometrics, 31,* 1–6).
15. Kim, W., Suh, S., & Han, J. J. (2015). Face liveness detection from a single image via diffusion speed model. *IEEE Transactions on Image Processing*, *24*(8), 2456–2465.
16. Lee, T. S. (1996). Image representation using 2D Gabor wavelets. *IEEE Transactions on Pattern Analysis and Machine Intelligence*, *18*(10), 959–971.
17. Määttä, J., Hadid, A., & Pietikäinen, M. (2012). Face spoofing detection from single images using texture and local shape analysis. *IET Biometrics*, *1*(1), 3–10.
18. Marasco, E., & Sansone, C. (2012). Combining perspiration-and morphology-based static features for fingerprint liveness detection. *Pattern Recognition Letters*, *33*(9), 1148–1156.
19. Menotti, D., Chiachia, G., Pinto, A., Schwartz, W. R., Pedrini, H., Falcao, A. X., & Rocha, A. (2015). Deep representations for iris, face, and fingerprint spoofing detection. *IEEE Transactions on Information Forensics and Security*, *10*(4), 864–879.
20. Nogueira, R. F., de Alencar Lotufo, R., & Machado, R. C. (2016). Fingerprint liveness detection using convolutional neural networks. *IEEE Transactions on Information Forensics and Security*, *11*(6), 1206–1213.
21. Ojala, T., Pietikainen, M., & Maenpaa, T. (2002). Multiresolution gray-scale and rotation invariant texture classification with local binary patterns. *IEEE Transactions on Pattern Analysis and Machine Intelligence*, *24*(7), 971–987.

22. de Freitas Pereira, T., Anjos, A., De Martino, J. M., & Marcel, S. (2013, June). Can face anti-spoofing countermeasures work in a real world scenario?. In *2013 International Conference on Biometrics, 12(1),* 1–8.
23. de Freitas Pereira, T., Komulainen, J., Anjos, A., De Martino, J. M., Hadid, A., Pietikäinen, M., & Marcel, S. (2014). Face liveness detection using dynamic texture. *EURASIP Journal on Image and Video Processing, 2014*(1), 1–15.
24. Phan, Q. T., Dang-Nguyen, D. T., Boato, G., & De Natale, F. G. (2016, September). FACE spoofing detection using LDP-TOP. In *2016 IEEE International Conference on Image Processing, 23(12),* 404–408.
25. Pinto, A., Pedrini, H., Schwartz, W. R., & Rocha, A. (2015). Face spoofing detection through visual codebooks of spectral temporal cubes. *IEEE Transactions on Image Processing, 24*(12), 4726–4740.
26. Pinto, A., Schwartz, W. R., Pedrini, H., & de Rezende Rocha, A. (2015). Using visual rhythms for detecting video-based facial spoof attacks. *IEEE Transactions on Information Forensics and Security, 10*(5), 1025–1038.
27. Poh, M. Z., McDuff, D. J., & Picard, R. W. (2010). Non-contact, automated cardiac pulse measurements using video imaging and blind source separation. *Optics Express, 18*(10), 10762–10774.
28. Sun, Z., Zhang, H., Tan, T., & Wang, J. (2013). Iris image classification based on hierarchical visual codebook. *IEEE Transactions on Pattern Analysis and Machine Intelligence, 36*(6), 1120–1133.
29. Taigman, Y., Yang, M., Ranzato, M. A., & Wolf, L. (2014). Deepface: Closing the gap to human-level performance in face verification. In *Proceedings of the IEEE Conference on Computer Vision and Pattern Recognition*, 81, 1701–1708.
30. Tronci, R., Muntoni, D., Fadda, G., Pili, M., Sirena, N., Murgia, G., ... & Roli, F. (2011, October). Fusion of multiple clues for photo-attack detection in face recognition systems. In *2011 International Joint Conference on Biometrics, 91,* 1–6.
31. Wen, D., Han, H., & Jain, A. K. (2015). Face spoof detection with image distortion analysis. *IEEE Transactions on Information Forensics and Security, 10*(4), 746–761.
32. Yadav, D., Kohli, N., Doyle, J. S., Singh, R., Vatsa, M., & Bowyer, K. W. (2014). Unraveling the effect of textured contact lenses on iris

recognition. *IEEE Transactions on Information Forensics and Security*, *9*(5), 851–862.

33. Rousson, M., Brox, T., & Deriche, R. (2003, June). Active unsupervised texture segmentation on a diffusion based feature space. In *2003 IEEE Computer Society Conference on Computer Vision and Pattern Recognition, 2003 Proceedings, 2*, 699.
34. Li, J., Wang, Y., Tan, T., & Jain, A. K. (2004, August). Live face detection based on the analysis of fourier spectra. In *Biometric Technology for Human Identification*, 5404, 296–303.
35. Pacut, A., & Czajka, A. (2006, October). Aliveness detection for iris biometrics. In *Proceedings 40th Annual 2006 International Carnahan Conference on Security Technology*, 12, 122–129.
36. Wei, Z., Qiu, X., Sun, Z., & Tan, T. (2008, December). Counterfeit iris detection based on texture analysis. In *2008 19th International Conference on Pattern Recognition*, 2(1), 1–4.
37. Tan, X., Li, Y., Liu, J., & Jiang, L. (2010, September). Face liveness detection from a single image with sparse low rank bilinear discriminative model. In *European Conference on Computer Vision*, 129, 504–517.
38. Peixoto, B., Michelassi, C., & Rocha, A. (2011, September). Face liveness detection under bad illumination conditions. In *2011 18th IEEE International Conference on Image Processing*, 34, 3557–3560.
39. Anjos, A., & Marcel, S. (2011, October). Counter-measures to photo attacks in face recognition: a public database and a baseline. In *2011 International Joint Conference on Biometrics, 3,* 1–7.
40. Schwartz, W. R., Rocha, A., & Pedrini, H. (2011, October). Face spoofing detection through partial least squares and low-level descriptors. In *2011 International Joint Conference on Biometrics, 31,* 1–8.
41. Zhang, L., Zhou, Z., & Li, H. (2012, September). Binary gabor pattern: An efficient and robust descriptor for texture classification. In *2012 19Th IEEE International Conference on Image Processing*, 5(3), 81–84.
42. Lee, T. W., Ju, G. H., Liu, H. S., & Wu, Y. S. (2013, May). Liveness detection using frequency entropy of image sequences. In *2013 IEEE International Conference on Acoustics, Speech and Signal Processing*, 31, 2367–2370.

43. Kose, N., & Dugelay, J. L. (2013, July). Reflectance analysis based countermeasure technique to detect face mask attacks. In *2013 18th International Conference on Digital Signal Processing, 12,* 1–6.
44. Doyle, J. S., Bowyer, K. W., & Flynn, P. J. (2013, September). Variation in accuracy of textured contact lens detection based on sensor and lens pattern. In *2013 IEEE Sixth International Conference on Biometrics: Theory, Applications and Systems, 7(1),* 1–7.
45. Erdogmus, N., & Marcel, S. (2013, September). Spoofing 2D face recognition systems with 3D masks. In *2013 International Conference of the BIOSIG Special Interest Group, 12,* 1–8.
46. Sequeira, A. F., Murari, J., & Cardoso, J. S. (2014, July). Iris liveness detection methods in the mobile biometrics scenario. In *2014 International Joint Conference on Neural Networks, 33,* 3002–3008.
47. Rattani, A., & Ross, A. (2014, September). Automatic adaptation of fingerprint liveness detector to new spoof materials. In *IEEE International Joint Conference on Biometrics*, *3*(1), 1–8.
48. Patel, K., Han, H., Jain, A. K., & Ott, G. (2015, May). Live face video vs. spoof face video: Use of moiré patterns to detect replay video attacks. In *2015 International Conference on Biometrics, 2(1),* 98–105.
49. Boulkenafet, Z., Komulainen, J., Feng, X., & Hadid, A. (2016, June). Scale space texture analysis for face anti-spoofing. In *2016 International Conference on Biometrics, 3(1),* 1–6.
50. Kumpituck, S., Li, D., Kunieda, H., & Isshiki, T. (2017, February). Fingerprint spoof detection using wavelet based local binary pattern. In *Eighth International Conference on Graphic and Image Processing, 10225*, 102251.

INDEX

Printed in the United States
by Baker & Taylor Publisher Services